Changing Representations of Nature and the City

"This ambitious publication makes an original contribution to our understanding of the changing perceptions of nature within the urban from the late 1960s/1970s. Architectural, visual and material culture case studies that include considerations of heritage, communication, and ecological perspectives from the UK, Europe, US and Asia contribute to its originality and offer new ways of thinking about and reflecting upon a major shift in contemporary global culture."

—*Fran Lloyd, Kingston University, UK*

The turn of the 1960s–1970s, characterized by the rapid acceleration of globalization, prompted a radical transformation in the perception of urban and natural environments. The urban revolution and related prospect of the total urbanization of the planet, in concert with rapid population growth and resource exploitation, instigated a surge in environmental awareness and activism. One implication of this moment is a growing recognition of the integration and interconnection of natural and urban entities. The present collection is an interdisciplinary inquiry into the changing modes of representation of nature in the city beginning from the turn of the 1960s/1970s. Bringing together a number of different disciplinary approaches, including architectural studies and aesthetics, heritage studies and economics, environmental science and communication, the collection reflects upon the changing perception of socio-natures in the context of increasing urban expansion and global interconnectedness as they are/were manifest in specific representations. Using cases studies from around the globe, the collection offers a historical and theoretical understanding of a paradigmatic shift whose material and symbolic legacies are still accompanying us in the early 21st century.

Gabriel N. Gee is Associate Professor in Art History at Franklin University, Switzerland.

Alison Vogelaar is Associate Professor of Communication and Media Studies at Franklin University, Switzerland.

Routledge Advances in Art and Visual Studies

This series is our home for innovative research in the fields of art and visual studies. It includes monographs and targeted edited collections that provide new insights into visual culture and art practice, theory, and research.

Perception and Agency in Shared Spaces of Contemporary Art
Edited by Cristina Albu and Dawna Schuld

Contemporary British Ceramics and the Influence of Sculpture
Iconoclasm, Monument, and Multiples
Laura Gray

Contemporary Citizenship, Art, and Visual Culture
Making and Being Made
Edited by Corey Dzenko and Theresa Avila

The Evolution of the Image
Political Action and the Digital Self
Edited by Marco Bohr and Basia Sliwinska

Artistic Visions of the Anthropocene North
Edited by Gry Hedin and Ann-Sofie N. Gremaud

Contemporary Artists Working Outside the City
Creative Retreat
Sarah Lowndes

Design and Visual Culture from the Bauhaus to Contemporary Art
Optical Deconstructions
Edited by Tóth

Changing Representations of Nature and the City
The 1960s–1970s and Their Legacies
Edited by Gabriel N. Gee and Alison Vogelaar

For a full list of titles in this series, please visit www.routledge.com/Routledge-Advances-in-Art-and-Visual-Studies/book-series/RAVS

Changing Representations of Nature and the City

The 1960s–1970s and Their Legacies

Edited by Gabriel N. Gee and Alison Vogelaar

Routledge
Taylor & Francis Group
LONDON AND NEW YORK

First published 2019 by Routledge

2 Park Square, Milton Park, Abingdon, Oxfordshire OX14 4RN
52 Vanderbilt Avenue, New York, NY 10017

Routledge is an imprint of the Taylor & Francis Group, an informa business

First issued in paperback 2020

Library of Congress Cataloging-in-Publication Data
A catalog record for this book has been requested

ISBN: 978-1-138-68853-7 (hbk)
ISBN: 978-0-367-58885-4 (pbk)

Typeset in Sabon
by Apex CoVantage, LLC

Contents

Figures

Acknowledgements

The idea for this publication began with a workshop in May 2014 hosted at Franklin University Switzerland by the TETI group—Textures and Experiences of Transindustriality. The workshop, entitled, "On the changing representation of nature in the trans-industrial city, 1960-present," brought together an interdisciplinary and international group of researchers to reflect upon transformations in urban landscapes over the past 50 years. The goal was to identify connections and commonalities in modes of representation of nature and cities as they had changed during that period. This publication is the result of our findings. The editors would like to thank the contributors for their commitment and willingness to pursue constructive dialogs over the past three years as well as our editor at Routledge, Felisa Salvago-Keyes. We would also like to thank Franklin University and the Swiss National Research Fund for their support in organizing the initial event. We hope this volume is just the beginning of a larger and much needed conversation on the topics developed here.

Cover Conor McFeely, Video Still From *The Weathermen Projects*, Sequence 2, 2012

Introduction

Global Cities, Global Nature

Gabriel N. Gee and Alison Vogelaar

Several significant developments in material cultures and social imaginaries point to the turn of the 1960s–1970s as nurturing a paradigmatic shift in the representation of urban and natural environments. The largest umbrella under which the present enquiry evolves is that of late 20th-century globalization processes. The term 'globalization' points to the 'compression' of the world in planetary terms. It emphasizes the presence and agency of a 'global scale' of organization, which can be understood as 'overriding' micro (regional) and meso (national, international) scales.[1] In effect however, the process of social, economic and cultural *integration* that globalization brings to the fore,[2] imply an *imbrication* of different scales—characterized in particular by the term 'glocalization'[3]—and actors—human and non—spiraling more or less willingly towards an ever greater *interconnected* spatial as well as temporal earthly environment. As such, globalization is not a new phenomenon. Although one can reach back to earlier global interactions and conceptualizations, the 'great opening up' brought about by the 'great discoveries' and European long distance maritime travels of the 15th- and 16th-century colonial pursuits stands as a seminal historical moment in the formation of an integrated world.[4] Suddenly the ocean void beyond the Pillars of Hercules materializes, and a radical transformation in the conception and organization of the earth occurs. On the planisphere designed by the Genovese cartographer Nicolaus de Caverio around 1506, the Portuguese mapping of the African coast extends to the Middle East, India and China, while to the West, the outline of the Eastern coast of the Americas appears: the sea could now be crossed over.[5] These discoveries mark the beginning of a decisive shaping of a homogeneous 'world space' bound by water rather than land.[6] Nineteenth-century industrialization processes extended this global space through the development of maritime and terrestrial routes, crossed by coal-fed ships, trains and intercontinental telegraphic cables, that facilitated the circulation of people, ideas, goods and capital.[7] The drastic acceleration of global fluxes initiated in the 1960s–1970s constituted both a continuation of earlier colonial and Enlightenment processes and imaginaries, and also a possible disruption of such processes and imaginaries.

The 1960s–1970s mark the beginning of what Roland Robertson (1992)[8] has described as the uncertainty phase of globalization; a period when the confluence of political, technological, scientific, economic and social transformations disrupted many established orders, radically *deterritorializing*, among other things, existing notions of place and its attendant imagined communities, identities and responsibilities. First introduced by Gilles Deleuze and Felix Guattari (1972)[9] to describe the fluid and fragmented nature of subjectivity in contemporary capitalistic culture, *deterritorialization* is a term that has been appropriated by scholars of globalization to describe the processes and implications of "the detachment of social and cultural practices from their ties to place."[10] This is not to suggest that location no longer matters but rather that "modern culture is less determined by location because location is increasingly penetrated by 'distance.'"[11] The local is therefore increasingly subject to non-local influences. While deterritorialization is not exclusive to this period of globalization, what makes it a "central force,"[12] is its pervasiveness and sheer banality;[13] even if one wished to live a truly 'local' life, the fact is, "the average daily life, in the context of globality, is shaped by structures, processes, and products that originate elsewhere."[14] Anthropologist Arjun Appadurai has described these deterritorialized experiences as 'disjunctures,' asserting that global flows of "people, machinery, money, images, and ideas now follow increasingly nonisomorphic paths." The shear speed, scale and volume of which has become so great that "disjunctures have become central to the politics of global culture."[15] Radical advances in transportation and media figure prominently into the processes of deterritorialization that characterize this period.

The introduction of the standardized shipping container in the maritime economy of the 1960s gave an unprecedented boost to the circulation of goods across the world, binding distant locations more than ever before.[16] In parallel, technological progress in air transportation shifted the mode of human's long distance travel from sea to air, with the adoption of jet airliners; suddenly, the roaming of the globe was no longer the lifetime adventure of Ibn Battuta (1304–1377),[17] the years-long exploration of Captain Cook (1728–1779)[18] or even the Victorian *tour de force* of Phileas Fogg's travel across the world in 80 days![19] In Michel Butor's poetic verses entitled '*Aerial network,*' the traveler in 'the plane number 8' seems to always remain above Bangkok, Karachi, Téhéran, and Athens, and in 'the plane number 9' of Honolulu, Los Angeles, Montréal; with Orly airport in Paris steadily on the horizon home.[20] With greatly democratized air transportation, mass tourism popularized the previously class-based *Grand Tour*, connecting formerly estranged places and people, and thereby 'intensifying' humans' experience of living within a global entity. There is also no underestimating the role of new media in such processes. Writing in the 1960s about televisual and electronic media, Canadian media scholar Marshall McLuhan proclaimed, "we have extended our central nervous system itself in a global embrace, abolishing both space and time as far as our planet is concerned."[21] While McLuhan focused prominently on the medium of the television and its psychological (i.e. hot versus cold media) and sociological effects (i.e. the global village), communication satellites represented an equally, if not more, important globalizing force. Before satellites made telecommunications global, the world was connected by "a few hundred poor-quality telephone circuits" made possible through submarine cable voice links.[22] That changed on June 20, 1969 when the Apollo 11 moon landing was broadcast on live TV to the very first worldwide audience. Though international TV broadcasts had been taking place since 1962 when the first telecommunication satellite, Telestar, connected parts of Western Europe and the United States of America, it was not until 1969 (one week before the lunar landing) that satellite service to the Indian Ocean region was made possible by the launch of the Intelsat III satellite.[23] While live global televisual events were revolutionary in the early 1970s, by the end of the decade, they had become commonplace. It was now 'natural' to exist in the same time and place as people half away across the globe. The last 50 years have in effect witnessed the emergence of a 'screen-sphere' (Vivian Sobchack), in which screens are not merely mediators of our knowledge of the world, but essential components of our existence.[24] There is no denying the impact of this upon our "sense of place" and related senses of being and belonging.[25]

One significant consequence of such processes has been the weakening of the nation-state as the go-to 'container' for social belonging and the articulation of sub-, supra- and post-nationalist forms of identification.[26] While the term 'globalization' did not come into parlance until the 1990s, increasing experiences of deterritorialization made possible by media and transportation technologies, have given rise to what has been called the 'global imaginary.' Borrowing Benedict Anderson's idea of the nation as an "imagined community" and echoing Roland Robertson's early definition of globalization as "the intensification of consciousness of the world as a whole,"[27] Charles Taylor has suggested that "the sense of how we all fit together" is increasingly becoming global.[28] In addition to, and perhaps more important than, the intensification and acceleration of global processes, the 1960–1970s saw the intensification of the *subjective awareness* of this new globality.[29] While neither media nor migration are new phenomena, and not all people have the means to traverse geo-social spaces, "the explosive expansion and commingling of media and migration are new."[30] As such, the work of the imagination now takes place in a global context where it has become, according to Appadurai, "a staging ground for action,"[31] and one that has, according to Manuel Castells, manifested in new forms of global civil society that allow communities to engage and resist global processes.[32]

Just as this period has been characterized by the rise of a global imaginary, it too is characterized by forms of retreat, differentiation and relocalization, especially in former colonies where processes of decolonization inspired forms of reterritorialization and differentiation surrounding issues of political and cultural belonging and representation. This period made visible the dark heritage and forgotten memories of the Atlantic slave trade,[33] during which time we see the fomentation of resistant narratives, individual and collective, inspired by the retreat of European colonial empires. The Biafran or Nigerian Civil War (1967–1970) is another poignant example of the complexities of postcolonial global transformations. That war can be traced back to the 1914 colonization and subsequent amalgamation of Northern and Southern Nigeria by the British as a matter of political convenience. Nigeria's independence in 1960, brought to the fore once more the economic, political and cultural differences disregarded, and in turn exacerbated, by this colonial amalgamation, inspiring a brutal civil war over territorial and political rights in Nigeria. Postcolonial processes of deterritorialization/reterritorialization were of course not unique to Nigeria and indeed characterized much of the political and cultural milieu during this period. These complexities also played out in former colonial 'centers,' where Europeans were seen in particular as reeling from the sudden realization of the mere 'regional validity' of their global cosmological expansion and conceptualization, and the fundamental differences that characterized the peripheries they had strived to colonize, educate and homogenize.[34]

This period also saw growing concerns about the new forms of imperialism inspired by global media forms and flows. UNESCO's *New World Information and Communication Order* (aka the MacBride Commission), first convened in 1977 as the *International Commission for the Study of Communication Problems*, articulated and explored these concerns manifesting in a 1981 report (the MacBride Report) that proposed a "new world communication order" designed to correct the unbalanced nature of media influence on the 'developing world' and promote equity in global media representations. While the report was strongly supported by most member nations, both the United Kingdom and the United States condemned the report, calling it an attack on the freedom of the press, and ultimately withdrew from UNESCO in protest (they later rejoined). Concerns about mass-media flows and influences in this period also manifested in the birth of the field of media studies,[35] which developed provocative lines of inquiry around the "soft power"[36] of Western media in the global economy and the rise of forms of "media imperialism."[37] The central themes of this volume, *nature* and the *city*, were sites that were indelibly influenced and shaped by the rising tensions associated with these global processes and forms. The following two sections briefly explore the significance of the rise of global cities and global natures during this period.

Global Cities: Urbanization and Industrialization

In 1970, the geographer Henri Lefebvre published *The Urban Revolution*, which effectively announced an ongoing process of *total urbanization* of the world, giving form to an all-encompassing 'urban society.'[38] Indeed, transformations of urban and natural landscapes have reached unprecedented levels in the past 40 years, leading observers to comment on the unstoppable expansion of urban entities. While metropoles counting more than a million inhabitants at the dawn of the 20th century could be counted on the fingers of one's hands, they are numbered in the hundreds at the

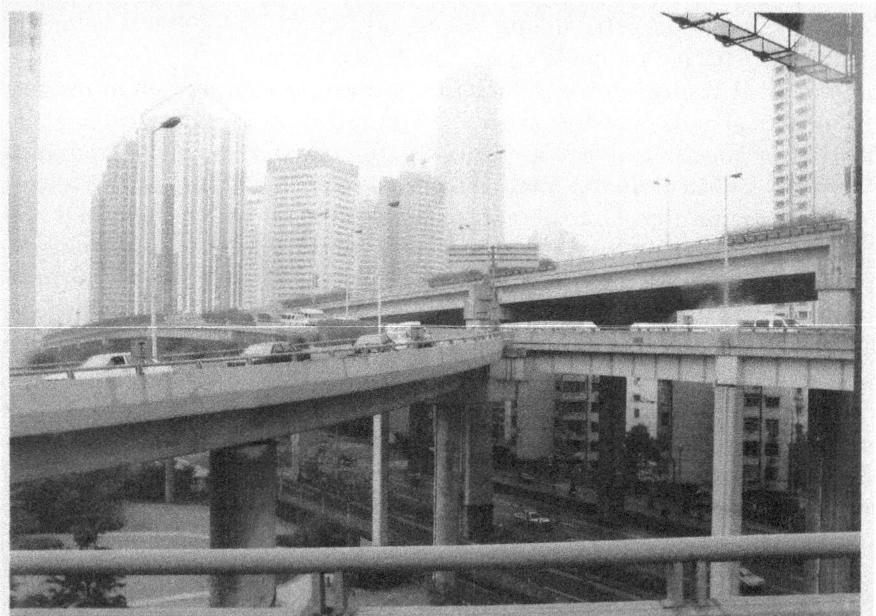

Figure 0.1 Anne-Laure Franchette, *Beijing*, 2005
courtesy of the artist

beginning of the 21st century.[39] This urban proliferation implies an inevitable reconfiguration of nature–city binaries. The writer Jean Giono in his evocation of the Valais region in Switzerland in the 19th century, captured the momentous change that had transformed mountainous environments:

> One has to imagine the mountain in 1850. It wasn't as it is today a ski factory with motorways, aeronautical fields, helicopters, chairlifts and skilifts, funiculars, hordes of cars and five star hotels. It was of a sordid heroism. No one went there, even in the summer, and those how stayed there in the winter, lived like marmots.[40]

In addition to the development of the tourism industries, an unstoppable process of urban sprawl was radically transforming the landscape of Alpine Switzerland.[41] Similarly, the 19th century witnessed a new appreciation of the seashore, crystallized by the invention of the beach. This novel sentiment was depicted by artists, from the early Gustave Courbet's pictorial salute to the sea in *Le bord de mer à Palavas* (1854), the chic promeneurs and resort of Claude Monet's *Hôtel des roches noirs* in Trouville Normandie (1870), via the morose mid-20th-century response to this burgeoning coastal sociology expressed by Cesare Pavese in his novel *The Beach* (1942),[42] right up to the 1980s fleshy sunbathers of the American painter Eric Fischl's beach series. Coasts have been privileged sites of human inhabitation and seaside civilizations for thousands of years.[43] However, the recent condensation of population on coasts has been accompanied by the emergence of larger, deeper and higher built environment. In New York, if the Swiss architect Le Corbusier moaned at what he saw as the city's chaotic urban planning (compared to his own as yet unrealized master plans!),[44] Serge Gainsbourg hanging on a construction beam in mid-air sang cooly impressed 'J'ai vu New York, haut c'est haut' (*I have seen New York, high it's so high*, 1964)![45] An immense horizontal watercolor by Christine Nicolas entitled *trait de côte* (2013, displayed on film through a 24.57mn traveling sequence) offers on the other hand one long suburban panorama of the Southern France coast near Marseille.[46] This regional coastal solidification was exemplarily witnessed on a national scale in Japan, which transformed its waterfront in one compact concrete interface during its 'high growth phase' in the 1960s and 1970s.[47] These vertical and horizontal processes of urban proliferation have been accompanied by the emergence of 'a transnational urban system,'[48] facilitating the cross-border circulation of materials and ideas, shaping truly global spaces whose 'transcultural' identities articulate novel forms of placemaking.[49]

Two key themes in the past 50 years are the emergence of 'megacities' and 'urban corridors,' trends which have effectively tied millions of people together through an unprecedented urban network. The urban revolution hasn't simply furthered the inner capacity of cities to produce and exchange wealth, it has also nurtured poverty. Mike Davis describes the outcome as a *planet of slums*, characterized by the formation of megacities and hypercities (the latter counting more than 20 million inhabitants, the equivalent as he points out to the estimated urban population in 1789), swallowing rural spaces and traditions, fomenting unequal urban development and hazardous urban living conditions.[50] This process is seen as creating new hybrid landscapes, neither urban nor rural, that have been shaped both by urban expansion and the urbanization of the countryside from within.[51] In this new global landscape, the stress shifts from urbanization, to *urbanizations*, the formation of multiple urban conditions, spaces and representations that transcend the former opposition between urban and non-urban/natural environments to define in Neil Brenner and Christian Schmidt's words an all-encompassing process of 'planetary urbanization.'[52] This phenomena must be seen in the light of the industrial revolutions of the 18th and 19th centuries, which prompted the advent of a modern age combining technological progress with 'industrial overflowings.'[53] Its linear narrative of progress was progressively disrupted by an awareness and sensitivity to environmental balance, risks and collapse, none more so than in the post WW2 era.[54] In this vein, the 'metabolic' processes explored by urban political ecology from the 1990s onwards have significantly underlined the imbrication of material and representational poles in the construction of our equally imbricated socio-natures.[55] It is the role of the cultural discursive and imaginary modes which convey meanings to the spatial-material processes of urban life, on which this volume places particular emphasis, at a turning point in the scale of urbanization processes.[56]

Furthermore, within this planetary turn, it is worth noting the additional virtual layers which have grown out of the information industry. In Melanie Jackson's animation film *A Global Positioning System* (2006), the narrative follows the global production line of 'killer mobiles,' the 'handheld satnav shootout,' from the military controlled diamond mines in sub-Saharan Africa to the anonymous computer production lines in Asia, ultimately reaching—via a fast postage service—the sofa of a consumer in the northern hemisphere. From the television screen to mobile portable internet devices, a texture and structure of electronic signals complements the all-defining planetarization,[57] whose physical and digital extensions articulate for Paolo Perulli a *Terra Mobile* requiring a reframing of our urban comprehension.[58] Late 20th-century globalization, urban revolution and planetary urbanization have triggered new ecological conditions, which have given rise to new and striking forms of environmental consciousness during the 1960s/70s.

Global Natures: Ecology and the Birth of an Environmental Imagination

On December 24, 1968, NASA's Apollo 8[59] was on its fourth orbit around the moon. Inside three astronauts, Frank Borman, James Lovell and William Anders, were busily collecting mission data. Though a 'man' would not land on the moon for another seven months, they were the first humans ever to leave the Earth's orbit and also the first to orbit around the moon. Just as the space shuttle nosed its way around the dark side of the moon, Lunar Module Pilot, William Anders, caught a view of the Earth rising over the moon's horizon from one of the small spacecraft windows. The mission audio tape recorded his awe: "Oh my God!" he said. "Look at that picture over there! There's the Earth coming up. Wow, is that pretty." Ander's awe did not remain his own. The crew broke protocol (the mission plan was to take photographs of the moon and not the Earth) and scrambled to snap several impromptu photos. Discontent with the first black and white images, the crew found color film and quickly snapped two more photos. One of these photos, *Earthrise*, quickly became the "choice of photo editors for the rest of history," and allegedly, "the most commonly published photograph in all of history."[60 61]

Earthrise has been mobilized over the past 60 years for a variety of causes from the funding of space programs and STEM fields to the promotion of world peace and environmentalism. In retrospect, it is clear why this image struck a chord in this

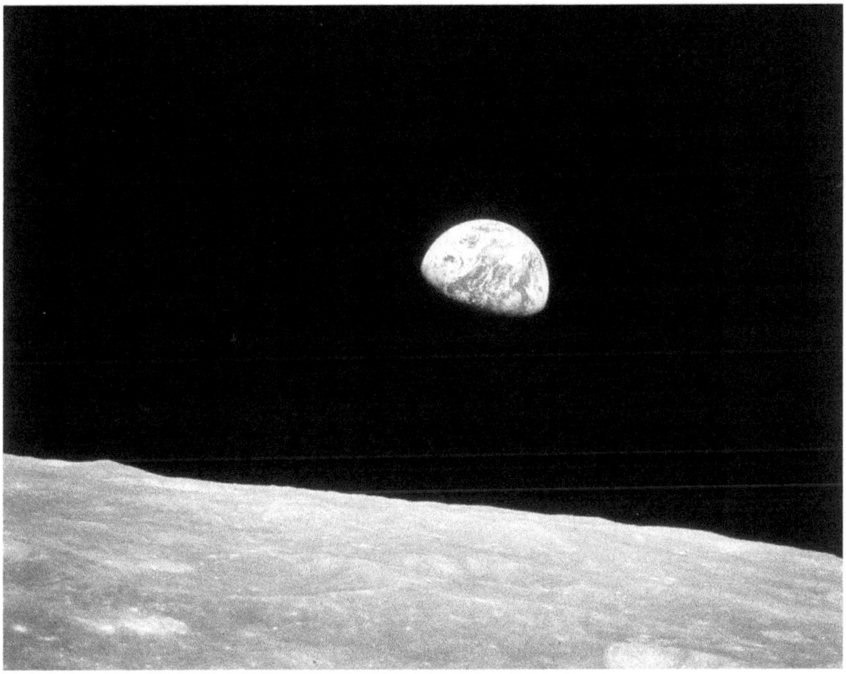

Figure 0.2 Bill Anders, *Earthrise*, 24.12.1968, NASA

moment. In capturing our vibrant, blue-white marble of a planet, hovering slightly above the horizon of a barren, brown-grey moon, *Earthrise* made visible the dreams and nightmares of life on a rapidly transforming planet. A planet rocked and reorganized by back-to-back world wars, the astounding and contradictory legacies of which included: radical advances in aerospace and chemical engineering, nuclear physics, computing and telecommunications, the symbolic and material investments in state-sponsored socialism and capitalism, the 'end' of European imperialism and subsequent processes of decolonization, and a drastic increase in the world's population attended by the global intensification of urbanization, consumerism and resource exploitation. It was, as Charles Dickens would describe an equally radical moment almost 200 years before, the best of times *and* the worst of times. It was the first time that a generation could describe the planet Earth (as opposed to a nation, region or dwelling) as home and the first time a generation could imagine that home being destroyed with the "press of a button." It was the birth of a large and affluent middle class in some parts of the world and the birth of a devastating cycle of dependency in others. It was a moment characterized by radical shifts in knowledge about natural and social worlds and a moment when that knowledge could be used to manipulate and exploit these worlds. It saw the development of technologies of truly global connectivity and the manipulation of these technologies in the service of 'free' markets.

Earthrise made visible (and indeed helped produce) the radical idea of 'global nature' emerging during this time period in scientific, philosophical and environmentalist circles. Prior to the advances in computer modeling, aerospace engineering and satellite communications that took place during this period, nature and environmental issues were understood as fundamentally local. These new modes of seeing and measuring helped make visible an integrated Earth system; one global nature in which damage to one part constituted damage to all parts. No group has used this image more pervasively than the American (and now global) environmental movement. Though its origins reach back into the late colonial period/Industrial Revolution, and it has roots in diverse movements, organizations and events,[62] many date the movement's birth with the 1962 publication of American marine biologist, Rachel Carson's *Silent Spring*,[63] a cautionary tale and scientific argument about the detrimental effects of synthetic pesticides on the environment.[64] Though she was vehemently (and predictably) criticized by the chemical industry for her scientific claims, and by some scientists for her literary style, the book was overwhelmingly supported by the scientific community and well received in popular and political spheres. *Silent Spring* was followed in 1968 by American population biologists Paul and Anne Ehrlich's book, *Population Bomb*,[65] which predicted catastrophic famines resulting from unchecked population growth. Though the book has been widely criticized for its unsubstantiated claims and incorrect predictions, it tapped into and reproduced cultural fears about overpopulation in that period. The Club of Rome's 1972 publication, *The Limits to Growth*,[66] rearticulated this fear, using new forms of computer-generated data sets and modeling, into the now pervasive constructs of "carrying capacity" and "systems collapse." In 1979, American scientist James Lovelock popularized his Gaia hypothesis, an idea he first developed while working for NASA at the Jet Propulsion Laboratory in California in the mid-1960s, with the publication of *Gaia: A New Look at Life on Earth*,[67] which attracted scientific and popular attention for its notion of a self-regulating Earth system. These discursive articulations of the risks associated with a 'global nature' were attended by legislation in many parts of the world although the development of global environmental governance did not officially take shape until much later in 1992 with the UN *Conference on Environment and Development's (UNCED)* Earth Summit in Rio De Janeiro. It is worth noting here the centrality of the environmental sciences in the early articulation of the environmental movement. Indeed, ecology and the environmental movement developed in tandem. Though its roots date back at least to 18th-century imperial science and botany, it was not until the 1960s that the interdisciplinary field of ecology gained public prominence. Environmental Historian Donald Worster has called this period, "the Age of Ecology," a period when the fate of the planet seemed to be in the hands of ecologists.[68] Both ecology and environmentalism are representative of the broader development and expansion of "systems thinking" in many domains and disciplines—an intellectual movement inspired in part by the new forms of imagining made possible by new technologies and in part by the daily experiences of life on a globalizing planet.

In spite of the many positive implications of articulating a 'global nature,' so much of the ecological thinking and environmental action in this period was firmly rooted in the Enlightenment nature–culture binary—a binary that incorrectly, and often unjustly, distinguished (some) culture from nature and designated the former as morally superior and the later as mere resource. Environmental research, policy and management often perpetuated this binary, albeit to the benefit of nature, studying and conserving only "wild" (non-human) nature[69] and ignoring environmental issues in densely populated urban areas where it was equally relevant and necessary. The articulation of *deep ecology* (by Norwegian philosopher Arne Naess), *political ecology* and *ecocriticism* in the early 1970s began a line of inquiry critical of the nature–culture binary and its underlying (Western) cultural and philosophical ideologies.[70] The interdisciplinary field of ecocriticism was particularly central in the identification and critical exploration of the nature–culture binary. One key study, Raymond Williams' 1973 book, *The Country and the City*, documented the changing (and often problematic) representations of the city and the country in British novels, noting the precedents for, and implications of, such representations on collective imaginations. The anthropological move 'beyond nature and culture' (Descola 2005) was accompanied by a crucial rearticulation and re-appreciation of different socio-natural systems throughout the world, thereby initiating a reframing of the history of Western Science as it developed from the 17th century onwards. As Descola put it: "the manner in which Western modernity represents nature is the least agreed upon thing in the world."[71] This rearticulation has led to the possibility of "decolonising nature" (Demos 2016),[72] and of rethinking the complex socio-political ties that link Western sciences to Western imperial expansions. Tellingly, in *What Plants Were Called Before They Had a Name* (2016–2017), the artist Uriel Orlow compiled an audio plant dictionary in nine indigenous South African languages.[73] This oral repository is to be seen in contrast to the Linnaen classification system used by the National Botanical Gardens of South Africa. A 're-imagining' of nature can benefit from an ecosemiotic attention to the merging of plural axis of meanings, from material to cultural, but also from 'deep' to 'dark' ecology.[74] Similarly, with respect to the nature–urban binary seen as an offshoot of the nature–culture binary, the recommendation of urban political ecology that urban milieus be considered as the product of socio-environmental processes, and thereby not unnatural per se, has further underlined the productive need to consider urban ecologies and ecosystems as an integral part of 'nature.'[75]

Likewise, a number of artistic practices emerging during the period can also be seen as shifting from and questioning modernist self-referential (anthropocentric) entrenchments to explore more ecological (ecocentric) attitudes and positions. Environment and Land Art of the 1960s and 1970s are prime examples of this epochal aesthetic transformation, which James Nisbet analyzes as part of a broader development in contemporary arts inducing a move towards an exploration of the complex interconnections binding global ecologies of information and power.[76] This engagement can be seen in the light of the period's aesthetic stretch beyond the traditional frames of modern art towards *expanded fields*.[77] The work of Robert Smithson for instance, has much to say about issues of 'localisation,' as exemplified by his site/non-sites series, and about trans-industrial processes in the second half of the 20th century, which imply equally complex superpositions of differently timed and located industrial cultures—his *Monuments of Pessaic* in Newark New Jersey is conceived as a vision of an entropic 'natural' disintegration of industrial textures, at a time of a potent restructuration in global economic infrastructures and cultures. Similarly, the artists associated with Arte Povera in Italy in the 1960s and 1970s can also be seen as emblematic in their manipulation of 'humble' materials, rich in spiritual and historical values at the time of the Italian 'economic miracle,' thereby articulating an alternate outlook in the face of a radiant society of consumption whose excesses and limits are now in urgent need of being controlled.[78] In parallel, the 1960s/70s opened pathways through which contemporary art could engage with the legacies of 'the age of empire,'[79] a condition Rasheed Araeen stressed concerns both former colonized and colonizers.[80] The emergence of a 'postcolonial constellation,' to use the terms of

Okwiu Enwezor, is by no means a smooth emancipatory revelation, but can highlight the conjunction of anti-hegemonic movements born from the political and cultural struggles taking place in the post-Second World War period.[81] Concomitant with the 1960s 'urban crisis' described by David Harvey in his evocation of 'rebel cities,' the period's civic unrests and grassroot activist protests channeled a political and philosophical query into the form and nature of cities in which communities want to live in: into their ecologies.[82]

In the last 20 years, distinct artistic and curatorial practices have given voice to these ecologic concerns stemming from society, such as the exhibition *Natural Reality: Kunstleriche Positionen zwischen Natur und Kultur in Aachen* (1999),[83] or *Radical Nature: Art and Architecture for a Changing Planet: 1969–2009* at the Barbican, London, which mapped key artistic proposals produced in response to environmental change since the 1960s (2009), and in the field of architecture events such as *Urban Nature in Megacities. Bi-Cities biennale of architecture* in Shenzen and Hong Kong (2014).[84] Eco-aesthetics, the title of an essay by Malcolm Miles published in 2014 which focused on artistic engagements with climate change, have blossomed, just as the concerns over the sheer acceleration of terrestrial environmental degradation have grown and voices from increasingly varied part of our global societies have risen to question the seemingly unstoppable logic of 'late capitalism.'[85] For example, a *Third Text* special issue from 2013 edited by T. J. Demos aimed to highlight a number of aesthetic strategies that engaged with the ecological crisis accentuated by the neoliberal turn of the 1980s; this collective discussion hence furthered a reflection on the modes through which artistic practices can intervene directly as agents in the politics of ecology.[86] For at the beginning of the 21st century, debates around 'the politics of nature' have taken an urgent turn, captured by the introduction of a new term to characterize a new era in the history of the earth's geology: the Anthropocene.[87] The word was coined to designate the present geological era, characterized by the impact of humans on the earth system. While the inception of such an influence can be traced to earlier historical landmarks (including that of the great discoveries which prompted the rapid disappearance of 50 million humans in the Americas), the analysis of parameters such as the atmospheric concentration in CO_2, the impoverishment of the ozone layer, the rise of temperature, of the disappearance of species, or of the loss in forest areas, all point to a starting point around 1800, and a rapid acceleration since 1945.[88] However, the term itself has come, tellingly, under critical scrutiny. As Heather Davies and Etienne Turpin underlined, the Anthropocene is not a mere descriptive word, but articulates a social-imaginary which potentially masks the heavy historical power relations that have prompted such a human impact on the earth.[89] Such is the need to weigh the political implications of the inherent metaphorical quality of scientific terminology,[90] that a range of alternate denominations has emerged, including the Plantationocene and the Capitalocene,[91] and the Chthulucene proposed by Donna Haraway,[92] all suggesting a critical turn 'against the Anthropocene.'[93] These debates highlight the emergence of oppositional forces intervening in the 'politics of nature,' which Bruno Latour sees as opening a path towards a new constitution, in which humans and non-humans are both given representational rights, in a necessary attempt to regulate and enforce a common earthy ecosystem.[94] We go back to the historical caesura of the 1960s/70s, when a certain idea of modernity rooted in the project of the Enlightenment came under scrutiny, when the intensity of urban life and urban proliferation accelerated, to reflect on the interconnectedness of socio-natures at a moment of exploding trans-industrial and transnational exchanges, which shaped the terms of the contemporary problem of the urban ecosystem.

The selected case studies in this volume, stemming from environmental studies and discourses, ecocriticism, socio-economics and political ecology, urban planning and architecture, critical heritage studies, history and art history, each contribute to defining, mapping and understanding *the unique and specific* characteristics of this historical change in modes of representing nature at the turn of the 1960s–1970s. This interdisciplinary exploration of the subject is therefore intentional. While a study focusing upon for instance on the modes of representing natural-urban components in the late 20th century in architectural practices would be very compelling; it would not, however, account for the emergence of *parallel* and *interconnected concerns* across disciplinary and social spheres. The object of this publication is precisely *to explore these commonalities*, and to reflect upon what the *representations* of this *historical shift in imagining nature* tell us about our present societies' constitution and challenges. For ultimately, the combined discussions are to contribute to the emergent queries in many different fields and activities about our present societies' struggle to implement socio-natural and culturally sustainable development, in the face of increasingly patent and threatening socio-natural collapses.

Additionally, the specific period considered (1960–1970s) is paramount to our study. In order to reflect upon the interconnected transformations that bind distant spaces during the period considered, the proposed chapters offer a number of case studies focusing on world-wide territories, as well as the passages between them. The volume considers different cultural and geographical representations, oscillating between chronologies anchored in European and North American histories, to perspectives stemming from Asian and Oriental histories, in order to draw a global understanding of the changing perception of nature in urban environments. Therefore on the one hand the collection includes a range of different disciplinary perspectives and objects, highlighting the paradigmatic character of the observed shift in natural and urban representations and imaginaries at the turn of the 1960s–1970s, and on the other hand it brings case studies from all continents, to reflect on its interconnected global spatial dimensions.

Forgotten Spaces, Artificial Spaces, Interstitial Spaces

The collection is organized in three sections, respectively underlining the issues of memory and loss, artificiality and simulacra, and hybrid and emancipated socio-natures central to this historical transformation. The first part, entitled 'Forgotten Spaces,' echoes the photographer Allan Sekula's 2010 documentary, *The Forgotten Space*, produced in collaboration with Noël Burch. For Sekula and Burch, the space in question is the sea, forgotten as a lived space full of meaningful symbolic and material associations, and transformed by economic forces into a mere channel for commodities across the globe. Chapters in this section consider in particular the forms of erasure produced by global shifts and trans-industrial mutations, and the memories that linger in the aftermath of landscape and habitat transformation. The first essay by Gabriel N. Gee builds out of Sekula and Burch's documentary project, exploring a series of artistic practices surrounding changing perceptions of maritime spaces that emerged at the turn of the 1960s–1970s. Using the mythological figure, Narcissus, Gee explores the evolution of port cities' imaginaries as they mirror the desires and fears of the civilizations and communities at their shores. In this essay, he notes specifically the ontological and representational shifts taking place in the 1960s–1970s, that saw port cities in the West collectively first forget and then rediscover their connections to the sea, and concludes with a preliminary exploration of these themes in the context of Taiwan. Our next essay picks up in nearby Bangkok where Worrasit Tantinipankul explores socio-natural transformations in Cold War era Bangkok. In this essay, Tantinipankul provides a historical account of the radical transformations to landscape and human–nature relations related to Bangkok's rapid industrialization and development under American influence in the 1970s. According to the author, this relationship was characterized by a profound and problematic suppression of nature and natural/traditional modes of living with and among Bangkok's complicated waterways and related landscapes. Our third contribution to this section, by Michelle Stefano, explores the socio-natures of Sparrows Point, a steel mill town outside Baltimore City, Maryland, USA that operated from the 1890s–1972. The essay, structured around a conversation with Elmer J. Hall, an author who grew up in the company town during the 1940s–1950s, explores the diverse, often hidden, interconnections and contrasts between industry and nature in this industrial site. Our final essay in this section, by Lori Gibbs, continues north to Newark, New Jersey, where the author explores the ways in which the metaphor, "the wasteland," came to symbolize New Jersey's socio-natural landscape both by insiders and outsiders. Gibbs traces the material and symbolic grounds on which this representation was constructed in the 1960s and how it has endured to this day in spite of several important and visionary architectural and landscape interventions that suggest that though New Jersey may be gritty, it is no mere "wasteland."

The second part focuses on artificial spaces, exploring in particular the conceptualization and *agency* of the idea of artifice. The pieces herein evoke *artificial spaces* not merely as oppositional to seemingly pristine, pure and authentic spaces, but rather as potent forms of representation. The chapters hence reflect upon the potency of the *artifice* in representations of nature and cities—and accordingly of the pendant idea of the *authentic* in nature. In the opening essay, Paolo Perulli reflects upon the profound changes in representations of both city and nature brought about by 1960s and 1970s global processes including planetary urbanization. *Passages*, the seminal text written by the German philosopher Walter Benjamin in the 1930s, serves as a starting point to his enquiry, in locating the modern origins of the post-second world war total urban expansion phenomena. In the postmodern age, Nature seems to have disappeared, subsisting as a mere trope evoked in the construction of man-made environments. Perulli's essay continues in the intellectual pursuit of new passages between the city, and a lost Nature, such that we might be able to reconfigure a much needed existential order. The question of landscape is at the heart of the following essay by Giuliano Sergio, who looks at transformations to the Italian landscape in the 1960s–1970s through the lense of the photographer Luigi Ghirri. In a territory laden with cultural heritage, Ghirri's neo-avant-garde photography participates in the invention of a new visual language that bridged the gap between archetypal representations of a pristine nature opposed to the artifice of human constructions, subverting the mythic opposition between countryside and the city, as well as the opposition between everydayness of life and the sacred aura of the work of art displayed in the museum. Sergio reflects on the photographer's creation of a 'theatre of landscape,' which encompasses composite architectures, objects and sites, in a celebration of place as seen and experienced by communities. In the following piece, the artist Tuula Närhinen discusses her artistic practice with Gabriel N. Gee, and in particular her uses of plastic waste collected on the shores of Finland. She also explains her interests in ecological history, in the figure of Ernst Haeckel, and scientific enquiry, which inform a practice that explores the thin borders between fact and fiction, and the surprising similarities between scientific and artistic processes. Artificial spaces are also explored by Ayala Levin through a study of the architectural and urban planning of Dodoma and Abuja, the respective capitals of Tanzania and Nigeria, designed and built from scratch in the 1970s. The essay focuses more specifically on the manner through which nature and natural resources were represented in the design of the two cities. The planning, realized by North American architecture firms, echoed both the decade's environmental turn in architectural planning in the US, and the new postcolonial spirit affirmed in sub-Saharan Africa. Levin also underlines the differences in both form and symbolic representation evident in the architectural planning and the use of green spaces, between oil rich Nigeria and agriculturally inclined Tanzania. Finally, we end this section with a text by the artist Uriel Orlow, written to accompany his piece *Grey, Green, Gold*, which explores the history of the South African flower, Strelitzia Regina, also known as *Mandela's Gold*. The flower was the object of artificial breeding in the 1960s, under colonial rule. Orlow's work and this essay explore the history and symbolic representation of the flower, in connection with the history of Nelson Mandela's incarceration on Robben Island and the garden he created there with his fellow prisoners. The artifice of colonial botanic is juxtaposed with the artificiality of colonial management, opposed by a resilient alternate vision.

Finally, the last section furthers an enquiry initiated by the group on Textures and Experiences of Transindustriality (TETI)[95] on "the construction of interstitial heritage," focusing in this volume more specifically on the *hybrid* and *interstitial* forms of perceiving and constructing nature that have emerged during this period, through the notion of 'interstitial spaces.' The section explores the idea of 'wilderness,' evoking a publication by Baptiste Lanaspeze entitled '*Marseille, Wild City*' that was accompanied by photographs by Geoffroy Mathieu.[96] Essays herein explore the city, and more generally the entangled city–nature interface, as a wild space, noting the emergence of alternative representations of nature and the urban during the period that do not posit a boundary between the two. The notion of interstice furthers our enquiry into liminal spaces, and the creation of spatial and mental alternatives to dominant forms of representational systems.

The first essay by Eliana Sousa Santos looks at British architectural historian Reyner Banham's writings from the 1960s to the 1980s. The redefinition of architecture as an expanded discipline in charge of designing fit environments for human activities was central to Banham thinking. The role of technology for such an ambition was paramount, and Santos describes the technocentric and ecocentric environmentalist trends that fueled the debates of the day. Santos focuses upon Banham's travels in Los Angeles, which the historian celebrated, against the prevalent opinion of the time, as the epitome of do-it-yourself resourcefulness and organic urban development. The essay ends in the desert, in a picturesque space, positioned between the beautiful and the sublime, where aesthetic contemplation come to supplant technological ambition. In the following chapter, Jundan Zhang explores the interstices in the representations of nature found in the writings of the Obscure Poetry group in China during the 1970s. Obscure Poetry developed in the context of the 'educated Youth Program,' which involved displacing urban educated youths to the countryside. Out of earshot of the urban social order, in contact with the rural world, the young poets who became central to this artistic movement found a space to develop their own interstitial voices. The historical context in which these works were created is significant, positioned at the end of the cultural revolution, and immediately before the shift in policies known as Open and Reform in 1978. Zhang offers a distinctly Chinese ecocritical reading of Obscure Poetry informed by Chinese traditions of socio-natures. The third chapter by Maria Joao Matos focuses on urban planning in Lisbon, Portugal, tracing the history of parkways and greenways in the city at the turn of the 1960s. Matos underlines the rising concerns with intensified urban sprawl, and the demand for increased access to green spaces. In Lisbon, she reflects on the history of the Park of Monsanto, a forest in the middle of a growing urban texture, whose relation with the city encapsulates changing perceptions of green spaces in Portugal. The creation of a 'green corridor' in the 1990s, linking the Park of Monsanto to the center of the city, furthers a possible model for a sustainable contemporary city through interstitial green alleys. The next text is a discussion between Northern Irish artist Conor McFeely and Gabriel N. Gee, taking as a starting point his *The Weathermen Projects*, a 2013–2014 work by McFeely that mixes references to meteorology, counter-revolutionary movements of the 1960s and historical tipping points. The interweaving of local and global forces is a recurrent topic in McFeely's work, as are scientific figures and scientific experimental processes. The discussion touches upon biologist Paul Kammerer's experiments on toads, as well as the Ganzfeld effect, both central to recent artistic explorations by McFeely. McFeely's work provides a significant insight into micro and macro cosmological representations, scientific imaginaries and (post)human nature in the second half of the 20th century. Finally, in our last chapter, Alison Vogelaar and Brack Hale examine the lingering effects of the complex colonial history of New Zealand as it has manifest in two significant urban natures: Christchurch Botanical Gardens and Riccarton Bush in Christchurch. The essay first reminds us of the centrality of the colonization of nature that accompanied European expansion across the globe, noting the ways in which the development of botanical gardens and natural reserves mirrored European territorial, cultural and corporeal exploitation and appropriation. Zooming into the 1970s, Vogelaar and Hale note the cultural, political and ecological postcolonial shifts towards the preservation and reinvention of native cultures and heritage in New Zealand. The essay concludes by noting the resolutely interstitial nature of Christchurch Botanical Garden and Riccarton Bush which serve both as custodians of the European settlers history, and experimental spaces in which to contemplate a pre-colonial New Zealand socio-nature. As such, they offer spaces of *negotiation* central to future the constitution and representation of diverse and novel socio-natures.

Notes

1 Saskia Sassen, *A Sociology of Globalization* (New York: Norton & Company, 2007), 11.
2 John Hartley, *Communication, Cultural and Media Studies: The Key Concepts* (London, New York: Routledge, 4th edition, 2011), 125–126.
3 Erik Swyngedouw, "Globalisation or 'Glocalisation'? Networks, Territories and Rescaling", *Cambridge Review of International Affairs* 17, no. 1 (2004), 25–48.
4 Jean-Michel Sallmann, *Le grand désenclavement du monde 1200–1600* (Paris: Payot & Rivages, 2011) Christian Grataloup, *Géohistoire de la mondialisation* (Paris: Arman Collin, 2007).

5 Christian Deluz, "La mer, infranchissable?" in *La mer, terreur et fascination*, eds. Alain Corbin and Hélène Richard (Paris: Bibliothèque Nationale de France, 2004), 29–37.
6 Peter Sloterdijk, *Spheres II. Globes* (South Pasadena: Semiotext), 783–818.
7 Lucien Laubier, "La révélation des merveilles de la mer," in *La mer, terreur et fascination*, eds. Alain Corbin and Hélène Richard (Paris: Bibliothèque Nationale de France, 2004), 149–164.
8 Roland Robertson, *Globalization: Social Theory and Global Culture* (New York: Sage, 2006).
9 Gilles Deleuze and Felix Guattari, *Anti-Oedipus: Capitalism and Schizophrenia* (London: Continuum, 2004).
10 Ursula Heise, *Sense of Place, Sense of Planet: The Environmental Imagination of the Global* (New York: New York University Press, 2008), 51.
11 John Tomlinson, "Globalization and Cultural Identity," in *The Global Transformations Reader: An Introduction to the Globalization Debate*, eds. D. Held and A. McGrew (London: Polity Press, 2003), 273.
12 Arjun Appadurai, "Disjuncture and Difference in the Global Cultural Economy," in *The Globalization Reader*, eds. F.J. Lechner and J. Boli (Australia: Blackwell Publishing, 2004).
13 John Tomlinson, *Globalization and Culture* (Chicago: University of Chicago Press, 1999).
14 Heise, *Sense*, 54.
15 Aryun Appadurai, "Disjuncture and Difference in the Global Cultural Economy," 105.
16 Martin Stopford, *Maritime Economics* (London, New York: Routledge, 2009, Third edition), 35–43.
17 Ibn Battuta, *Voyages. 1 De l'Afrique du Nord à la Mecque* (Paris: La Découverte, 1997).
18 W.J.L. Wharton (ed.), *Captain Cook's Journal During His First Voyage Round the World Made in H.M. Bark 'Endeavour', 1768–71* (London: E Stock, 1893).
19 Jules Verne, *Le tour du monde en quatre-vingt jours* (Paris: Hetzel, 1873).
20 Michel Butor, "Réseau aérien" (1962), in *Anthologie nomade*, ed. Michel Butor (Paris: Gallimard, 2004).
21 Marshall McLuhan, *Understanding Media: Extensions of Man*, edited by W Terrence Gordon (California: Gingko Press, 2003), 5.
22 Joseph N. Pelton, "Satellites as Worldwide Change Agents," in *International Communication: A Reader*, ed. D.K Thussu (New York: Routledge, 2010), 17.
23 Pelton, *Satellites*.
24 Vivian Sobchack, "Comprendre les écrans: une mediation *in medias res*," in *Vivre par(mi) les écrans*, eds. Mauro Carbone, Anna Caterina Dalmasso, and Jacopo Bodini (Dijon: Les presses du Réel, 2016), 29–46.
25 Heise, *Sense*.
26 Sheila L. Croucher, *Globalization and Belonging: The Politics of Identity in a Changing World* (Oxford: Rowman and Littlefield Publishers, 2004).
27 Robertson, *Globalization*, 8.
28 Charles Taylor, *Modern Social Imaginaries* (North Carolina: Duke University Press, 2004), 24.
29 Manfred B. Steger, *The Rise of the Global Imaginary: Political Ideologies from the French Revolution to the Global War on Terror* (New York: Oxford University Press, 2008) 10–11.
30 Jack Lule, *Globalization and Media: Global Village of Babel* (Maryland: Rowman and Littlefield, 2015), 71.
31 Arjun Appadurai, *Modernity at Large: Cultural Dimensions of Globalization* (Minneapolis: University of Minnesota Press, 1996), 22.
32 Manuel Castells, "The New Public Sphere: Global Civil Society, Communication Networks, and Global Governance," in *International Communication: A Reader*, ed. DK Thussu (New York: Routledge, 2010).
33 Paul Gilroy, *The Black Atlantic: Modernity and Double Consciousness* (London: Verso, 1993).
34 Peter Sloterdijk, *Spheres II. Globes*, 749–750.
35 First called International Communication Studies.
36 See for example Joseph S. Nye Jr., "Public Diplomacy and Soft Power," in *International Communication: A Reader*, ed. D.K Thussu (New York: Routledge, 2010).

37 See for example Oliver Boyd-Barrett, "Media Imperialism: Towards an International Framework for the Analysis of Media Systems," *Mass Communication and Society* (1977), 116–135.
38 Henri Lefebvre, *La révolution urbaine* (Paris: Gallimard, 1970).
39 David Harvey, "Cities or Urbanization?" in Neil Brenner, ed. *Implosions/Explosions. Towards a Study of Planetary Urbanization* (Berlin: Verlag, 2014), 52–66.
40 Jean Giono, *Le déserteur* (Paris: Gallimard, 1973), 34: "Qu'on se représente la montagne en 1850. Ce n'était pas comme maintenant une usine à skis avec autoroutes, champs d'aviation, hélicopters, télésièges, remonte-pentes, funiculaires, caravanes de cars et hôtels vingt-cinq étoiles. C'était l'héroïsme sordide. On n'y venait pas, même en été, et ceux qui l'habitaient l'hiver y résidaient en marmottes" (authors translation).
41 Roger Diener and all (eds.), *La Suisse: portrait urbain. Volume 1 Introduction; Volume 2 Frontières communes: brève histoire du territoire Volume 3 Matériaux* (Basel, Boston, Berlin: Birkhäuser, 2005); see also Christian Schmid, "Travelling Warrior and Complete Urbanization in Switzerland: Landscape as Lived Space," in *Implosions/Explosions: Towards a Study of Planetary Urbanization*, ed. Neil Brenner (Berlin: Jovis Verlag, 2014), 90–108. In this revealing essay on the urbanization of Switzerland, Schmid takes as his starting point a 1979 documentary film by Christian Schocher, in which the audience is taken onto a car journey throughout the country, depicting the extent to which Switzerland has become an 'urban landscape.'
42 Cesare Pavese, *La spiaggia* (Turin: Einaudi, 1942).
43 John R. Gillis, *The Human Shore* (Chicago: The University of Chicago Press, 2012).
44 Rem Koolhas, *New York délire* (Marseille: Parenthèse, 2002), 256–259.
45 Serge Gainsbourg, *New York USA* (1964).
46 Exhibited in Marseille at the exhibition in 2013 *J'aime les panoramas* (Marseille: Mucem, 2015).
47 Augustin Bercque, *Les sauvages et l'artifice. Les japonais devant la nature* (Paris: Gallimard, 1986).
48 Saskia Sassen, A Sociology of Globalization, 28–30.
49 Jeffrey Hou (ed.), *Transcultural Cities: Border Crossings and Placemaking* (London, New York: Routledge, 2013).
50 Mike Davis, *Planet of Slums* (London: New York, Verso, 2006).
51 Davis, *Planet of Slums*, 7–10. The references are in particular to Gregory Guldin, *What's a Peasant to Do? Village Becoming Towns in Southern China* (Boulder and Oxford: Westview Press, 2001). Also evoked are Tomas Sieverts notion of the 'Zwischenstadt,' characterized by decentralized urban structures in *Cities Without Cities: An Interpretation of the Zwischenstadt* (London: Taylor and Francis, 2003), David Drakakis-Smith blurred urban boundaries explored in *Third World Cities* (London, New York: Routledge, 2000) and the Indonesian Desakota regions, a hybrid of rural and urban spaces studied by T.G. McGee, in Norton Ginsburg, Bruce Koppel, and T.C. McGee, *The Extended Metropolis: Settlement Transition in Asia* (Honolulu: University of Hawaii Press, 1991).
52 Neil Brenner and Christian Schmidt, "Planetary Urbanization," in *Implosions/Explosions: Towards a Study of Planetary Urbanization*, ed. Neil Brenner (Berlin: Jovis Verlag, 2014), 160–163.
53 Thomas le Roux and Michel Letté (eds.), *Débordements industriels. Environnements, territoire et conflit XVIII-XXI siècles* (Rennes: Presses Universitaires de Rennes, 2013).
54 Ulrich Beck, *La société du risque. Sur la voie d'une autre modernité* (Paris: Flammarion, 2001, first published in Frankfurt am Main, Verlag, 1986).
55 Erik Swyngedouw, *Social Power and the Urbanization of Water: Flows of Power* (Oxford: Oxford University Press, 2004); Nick Heynen, Maria Kaika and Erik Swyngedouw, *In the Nature of Cities: Urban Political Ecology and the Politics of Urban Metabolism* (London: Taylor and Francis, 2006).
56 Christopher Schliephake in *Urban Ecologies: City Space, Material Agency and Environmental Politics in Contemporary Culture* (London: Lextington Book, 2015) underlines that "although (. . .) spatial-material processes constitute the framework of urban life, it is on the cultural-discourse level that their inner workings and interrelations are imbued with meanings," 12. On a parallel note, the art historian Pieter Krieger has outlined the need to

develop a strategic research at the crossroad of art history and anthropology on the urban imaginaries fomented by Megacities, see also Peter Krieger, "Aesthetics and Anthropology of Megacities. A New Field of Art Historical Research," in *Cannibalismes disciplinaires: quand l'histoire de l'art et l'anthropologie se rencontrent*, eds. Thierry Dufrêne and Anne-Christine Taylor (Paris: coédition INHA / musée du quai Branly, 2009), 197–212.
57 Massimo di Felice, *Paesaggi post urbani. La fine dell'esperienza urbana e le forme communicative dell'abitare* (Milano, Roma: Bevino, 2010).
58 Paolo Perulli, *Terra mobile* (Turin: Einaudi, 2014).
59 Apollo 8 is the eighth mission in NASA's race to put a man on the moon.
60 Christopher Riley, "Happy Birthday Earthrise," *BBC News*, December 24, 2008, retrieved on August 10, 2016 from http://news.bbc.co.uk/go/pr/fr/-/2/hi/science/nature/7797439.stm.
61 Al Gore A. and Davis Guggenheim, *An Inconvenient Truth* (Paramount Classics, 2006).
62 For example, Romanticism in Europe, the creation of National Parks in the United States, conservation societies in India, the United Kingdom and the United States, preservation societies in much of the colonial world, and environmental pollution and legislation in the US and the UK.
63 Rachel Carsons, *Silent Spring* (Boston: Houghton Mifflin, 1962).
64 Greg Garrard, *Ecocriticism* (New York: Routledge, 2011).
65 Paul R. and Anne Ehrlich, *Population Bomb* (New York: Sierra Club/Ballantine Books, 1972).
66 Donatella H. Meadows, Dennis L. Meadows, Jørgen Randers, and William. W Behrens. *Club of Rome: The Limits to Growth. A Report for the Club of Rome's Project on the Predicament of Mankind* (New York: Universe Books, 1972).
67 James Lovelock, *Gaia: A New Look at Life on Earth* (Oxford: Oxford University Press, 1979).
68 Donald Worster, *Nature's Economy: A History of Ecological Ideas* (Cambridge: Cambridge University Press, 1994). The emergence and proliferation of television documentaries on wild flora and fauna in the 1960s is also exemplary of and conducive to an increased public ecological awareness; see Keith Scholey, "A History of Half a Century of Wildlife Television and Its Impact on Audiences," in *Changing Perceptions of Nature*, eds. Ian Convery and Peter Davis (Martlesham: Boydell Press, 2016).
69 Paul Robbins, *Political Ecology: A Critical Introduction* (Hoboken: John Wiley and Sons, 2011).
70 Alfred K. Siewers, *Re-Imagining Nature: Environmental Humanities and Ecosemiotics* (Plymouth: Bucknell University Press, 2013).
71 Philippe Descola, *Par dela nature et culture* (Paris: Gallimard, 2005) 70, 'la manière dont l'occident moderne se représente la nature est la chose du monde la moins bien partagée.'
72 T.J. Demos, *Decolonizing Nature: Contemporary Art and the Politics of Ecology* (Berlin: Sternberg Press, 2016).
73 *What Plants Were Called Before They Had a Name* (2016–17) was shown at Corner College, Zurich, as part of Uriel Orlow exhibition entitled *Geraniums Are Never Red*. See also "Interview with Uriel Orlow", by Dimitrina Sevova (2017): <http://materials.corner-college.com/2017/201704/interview-uriel-orlow-by-dimitrina-sevova.html>
74 Siewers (ed.), *Re-Imagining Nature.*
75 David Harvey, *Justice, Nature and the Geography of Difference* (Oxford: Blackwell, 1996); Erik Swyngedouw, "Circulations and Metabolisms. (Hybrid) Natures and (Cyborg) Cities." *Science as Culture* 15, no. 2 (June 2006), 105–121; Nik Heynen, "Urban Political Ecology I: The Urban Century," *Progress in Human Geography*, 38, no. 4 (2014).
76 James Nisbet, *Ecologies, Environments, and Energy Systems in the Art of the 1960s and 1970s* (Cambridge, MA; London: MIT Press, 2014), 12.
77 A point made by Malcolm Miles in *Eco-Aesthetics. Art, Literature and Architecture in a Period of Climate Change* (2014), echoing the seminal essay by Rosalind Krauss, "Sculpture in the Expanded Field", *October* 8 (Spring 1979), 30–44.
78 Giovanni Lista, *Arte Povera* (Milano: 5 continents, 2006).
79 E.J. Hobsbawm, *The Age of Empire* (London: Weidenfield and Nicholson, 1987).
80 Rasheed Araeen, "Art and Postcolonial Society," in *Globalization and Contemporary Art*, ed. Jonathan Harris (Chichester: Willey-Blackwell, 2011), 365–374.

81 Okwui Enwezor, "The Postcolonial Constellation. Contemporary Art in a State of Permanent Transition," in *Fault Lines. Contemporary African Art and Shifting Landscapes*, eds. Gliane Tawadros and Sarah Campbell (London: Iniva, 2003), 65–79.
82 David Harvey, *Rebel Cities: From the Right to the City to the Urban Revolution* (New York: Verso, 2012).
83 *Natural Reality: Kunstleriche Positionen zwischen Natur und Kultur in Aachen* (Stuttgart: DACO, 1999), curated by Heike Strelow.
84 *Radical Nature: Art and Architecture for a Changing Planet: 1969–2009* (Köln: Walther König, 2010).
85 Fredric Jameson, "Postmodernism, or the Cultural Logic of Late Capitalism," *New Left Review*, n. 146 (July–August 1984), 53–92.
86 T.J. Demos, "Contemporary Art and the Politics of Ecology: An Introduction," *Third Text*, 27, no. 1 (January 2013), 1–9.
87 Christophe Bonneuil and Jean-Baptiste Fressoz, *L'événement anthropocène. La terre, l'histoire et nous* (Paris: Le Seuil, 2016). The term was first proposed by the chemist Paul Crutzen in a geosphere-biosphere conference in Mexico in 2000, which he subsequently theorized in the article "Geology of mankind", *Nature*, 415 (January 2002).
88 Christophe Bonneuil and Jean-Baptiste Fressoz, *L'événement anthropocène. La terre, l'histoire et nous*, 24–25.
89 Heather Davis and Etienne Turpin, *Art in the Anthropocene. Encounters Among Aesthetics, Politics, Environments and Epistemologies* (London: Open humanities Press, 2015).
90 Brendon Larson, *Metaphors for Environmental Sustainability: Redefining Our Relationship With Nature* (New Haven: Yale University Press, 2011).
91 Jason Moore (ed.), *Anthropocene or Capitalocene? Nature, History, and the Crisis of Capitalism* (Oakland: PM Press, 2016).
92 Donna Haraway, "Anthropocene, Capitalocene, Plantationocene, Chthulucene: Making Kin," *Environmental Humanities*, 6 (2015), 159–165: "I also insist that we need a name for the dynamic ongoing sym-chthonic forces and power of which people are a part, within which ongoingness is at stake. Maybe, but only maybe, and only with intense commitment and collaborative work and play with other terrans, flourishing for rich species assemblages that include people will be possible. I am calling all this the Chthulucene—past, present and to come."
93 T.J. Demos, *Against the Anthropocene*, conference held at Zurich University of the Arts on 07.04.2017.
94 The need to establish a new political contract that would go beyond human constituencies was advocated by Michel Serres in *Le contrat naturel* (Paris: Flammarion, 1992), and developed by Bruno Latour in *Politiques de la nature. Comment faire entrer les sciences en démocratie?* (Paris: La découverte, 2004) and *Face à Gaïa. Huit conférences sur le nouveau régimes climatique* (Paris: La Découverte, 2015). Reflections delving in political philosophy have multiplied in the past 30 years, from Hans Jonas, *Das Prinzip Verantwortung* (Frankfurt: Insel Verlag, 1979), to recent publications such as Catherine and Raphael Larrère, *Du bon usage de la nature. Pour une philosophie de l'environnement* (Paris: Aubier, 1997) or Dominique Bourg and Kerry Whiteside, *Vers une démocratie écologique. Le citoyen, le savant et le politique* (Paris: le Seuil, La république des idées, 2010).
95 <www.tetigroup.org>
96 Baptiste Lanaspeze and Geoffroy Mathieu, *Marseille Ville sauvage* (Aix-en Provence: Actes Sud, 2012).

References

Appadurai, Arjun. *Modernity at Large: Cultural Dimensions of Globalization*. Vol. 1. Minneapolis: University of Minnesota Press, 1996.

Appadurai, Arjun. "Disjuncture and Difference in the Global Cultural Economy." in F.J. Lechner and J. Boli (eds.), *The Globalization Reader*. Australia: Blackwell Publishing, 2004.

Araeen, Rasheed. "Art and Postcolonial Society." in Jonathan Harris (ed.), *Globalization and Contemporary Art*, 365–374. Chichester: Willey-Blackwell, 2011.

Battuta, Inn. *Voyages. 1 De l'Afrique du Nord à la Mecque*. Paris: La Découverte, 1997.
Beck, Ulrich. *La société du risque. Sur la voie d'une autre modernité*. Paris: Flammarion, 2001.
Bercque, Augustin. *Les sauvages et l'artifice. Les japonais devant la nature*. Paris: Gallimard, 1986.
Bonneuil, Christophe, and Jean-Baptiste Fressoz. *L'événement anthropocène: la Terre, l'histoire et nous*. Paris: Seuil, 2013.
Bourg, Dominique, and Kerry Whiteside. *Vers une démocratie écologique*. Paris: Editions du Seuil et la République des Idées, 2010.
Boyd-Barrett, Oliver. "Media Imperialism: Towards an International Framework for the Analysis of Media Systems." *Mass Communication and Society* (1977): 116–135.
Brenner, Neil, and Christian Schmidt. "Planetary Urbanization." in Neil Brenner (ed.), *Implosions/Explosions: Towards a Study of Planetary Urbanization*, 160–163. Berlin: Jovis Verlag, 2014
Brown, Robert MW. "Anthropocene or Capitalocene? Nature, History, and the Crisis of Capitalism edited by Jason W. Moore." *The Goose* 15, no. 2 (2017): 11.
Butor, Michael. "Réseau aérien" (1962). in Michel Butor (ed.), *Anthologie nomade*. Paris: Gallimard, 2004.
Carson, Rachel. *Silent Spring*. Boston: Houghton Mifflin Harcourt, 2002.
Castells, Manuel. "The New Public Sphere: Global Civil Society, Communication Networks, and Global Governance." in D.K Thussu (ed.), *International Communication: A Reader*, 36–47. New York: Routledge, 2010.
Convery, Ian, and Peter Davis (eds.). *Changing Perceptions of Nature*. Martlesham: Boydell Press, 2016.
Croucher, Sheila. *Globalization and Belonging: The Politics of Identity in a Changing World*. Oxford: Rowman & Littlefield Publishers, 2003.
Crutzen, P.J. "Geology of Mankind." *Nature* 415 (2002): 23.
Davis, Heather, and Etienne Turpin. *Art in the Anthropocene: Encounters Among Aesthetics, Politics, Environments and Epistemologies*. London: Open Humanities Press, 2015.
Davis, Mike. *Planet of Slums (2006)*. London & New York: Verso Google Scholar, 2007.
Di Felice, Massimo. *Peasaggi post-urbani: la fine dell'esperienza urbana e le forme comunicative dell'abitare*. Milano: Bevivino, 2010.
Deleuze, Gilles, and Felix Guattari. *Capitalism and Schizophrenia, Volume 1: Anti-Oedipus*. Mark Seam, Robert Hurley, and Helen Lane (trans.), London: Continuum, 2004.
Deluz, Christian. "La mer, infranchissable?" in Alain Corbin and Hélène Richard (eds.), *La mer, terreur et fascination*, 29–37. Paris: Bibliothèque Nationale de France, 2004.
Demos, T.J. *Decolonizing Nature: Contemporary Art and the Politics of Ecology*. Berlin: Sternberg Press, 2016.
Demos, T.J. "Contemporary Art and the Politics of Ecology: An Introduction." *Third Text* 27, no. 1 (January 2013): 1–9.
Diener, Roger, and all (eds.). *La Suisse: portrait urbain. Volume 1 Introduction; Volume 2 Frontières communes: brève histoire du territoire Volume 3 Matériaux*. Basel, Boston, Berlin: Birkhäuser, 2005.
Enwezor, Okwui. "The Postcolonial Constellation: Contemporary Art in a State of Permanent Transition." *Research in African Literatures* 34, no. 4 (2003): 57–82.
Erlich, Paul R., and Anne Ehrlich. *Population Bomb*. New York: Sierra Club/Ballantine Books, 1972.
Garrard, Greg. *Ecocriticism*. New York: Routledge, 2011.
Gillis, John R. *The Human Shore: Seacoasts in History*. Chicago: University of Chicago Press, 2012.
Gilroy, Paul. *The Black Atlantic: Modernity and Double Consciousness*. Cambridge: Harvard University Press, 1993.
Gore, Al, and Davis Guggenheim. *An Inconvenient Truth*. Paramount Classics, 2006.

Haraway, Donna. "Anthropocene, Capitalocene, Plantationocene, Chthulucene: Making Kin." *Environmental Humanities* 6, no. 1 (2015): 159–165.

Harris, Jonathan (ed.). *Globalization and Contemporary Art*. Chichester: Willey-Blackwell, 2011.

Hartley, John. *Communication, Cultural and Media Studies: The Key Concepts*. London: Routledge, 2012.

Harvey, David, and Bruce Braun. *Justice, Nature and the Geography of Difference*. Oxford: Blackwell, 1996.

Harvey, David. *Rebel Cities: From the Right to the City to the Urban Revolution*. London: Verso Books, 2012.

Heise, Ursula K. *Sense of Place and Sense of Planet: The Environmental Imagination of the Global*. Oxford: Oxford University Press, 2008.

Heynen, Nikolas C., Maria Kaika, and Erik Swyngedouw. *In the Nature of Cities: Urban Political Ecology and the Politics of Urban Metabolism*. Vol. 3. Oxon: Taylor & Francis, 2006.

Heynen, Nik. "Urban Political Ecology I: The Urban Century." *Progress in Human Geography* 38, no. 4 (2014): 598–604.

Hobsbawm, Eric J. "The Age of Empire, 1875–1914 (London: Weidenfeld and Nicolson, 1987)." in *Trad. Française: L'ère des empires,* Paris: *Fayard*, 253. 1989.

Hou, Jeffrey (ed.). *Transcultural Cities: Border Crossings and Placemaking*. London, New York: Routledge, 2013.

Jameson, Frederic. "Postmodernism, or the Cultural Logic of Late Capitalism." *New Left Review*, no. 146 (July-August 1984): 53–92.

Jonas, Hans. *Das Prinzip Verantwortung*. Frankfurt: Insel Verlag, 1979.

Krauss, Rosalind. "Sculpture in the Expanded Field." *October* 8 (1979): 31–44.

Krieger, Peter. "Aesthetics and Anthropology of Megacities: A New Field of Art Historical Research." in Thierry Dufrêne and Anne-Christine Taylor (eds.), *Cannibalismes disciplinaires: quand l'histoire de l'art et l'anthropologie se rencontrent*, 197–212. Paris: coédition INHA / musée du quai Branly, 2009.

Koolhaas, Rem. *New York délire*, 256–259. Marseille: Parenthèse, 2002.

Lanaspeze, Baptiste, and Geoffroy Mathieu. *Marseille Ville sauvage*. Aix-en Provence: Actes Sud, 2012.

Larrère, Catherine, and Raphaël Larrère. *Du bon usage de la nature: pour une philosophie de l'environnement*. Paris: Aubier, 1997.

Larson, Brendon. *Metaphors for Environmental Sustainability: Redefining Our Relationship With Nature*. New Haven: Yale University Press, 2011.

Latour, Bruno. *Face à Gaïa: huit conférences sur le nouveau régime climatique*. La découverte, 2015.

Latour, Bruno. *Politiques de la nature: comment faire entrer les sciences en démocratie*. Paris: La découverte, 2016.

Laubier, Lucien. "La révélation des merveilles de la mer." in Alain Corbin and Hélène Richard, (eds.), *La mer, terreur et fascination*. Paris: Bibliothèque Nationale de France, 2004.

Lefebvre, Henri. *La révolution urbaine*. Paris: Gallimard, 1970.

Le Roux, Thomas, and Thomas Letté. *Débordements industriels: environnement, territoire et conflit:(XVIIIe-XXIe siècle)*. Rennes: Presses universitaires de Rennes, 2013.

Lista, Giovanni. *Arte Povera*. Milano: 5 continents, 2006.

Lule, Jack. *Globalization and Media: Global Village of Babel*. Maryland: Rowman & Littlefield, 2012.

Lovelock, James, and James E. Lovelock. *Gaia: A New Look at Life on Earth*. Oxford: Oxford Paperbacks, 2000.

Malcolm Miles, "Eco-Aesthetics: Art, Literature and Architecture in a Period of Climate Change." *Environmental Philosophy* 12, no. 2 (2015): 277–280.

Malcolm Miles, *Eco-aesthetics: art, literature and architecture in a period of climate change*, London: Bloomsbury, 2014

McLuhan, Marshal. *Understanding Media: Extensions of Man*, W. Terrence Gordon (ed.). California: Gingko Press, 2003.
Meadows Donella, H., Dennis L. Meadows, Jorgen Randers, and William W. Behrens III. *The Limits to Growth: A Report for the Club of Rome's Project on the Predicament of Mankind*, 208. New York: Universe Books, 1972.
Mucem. *J'aime les panoramas*. Marseille: Mucem, 2015.
Moore, Jason (ed.). *Anthropocene or Capitalocene? Nature, History, and the Crisis of Capitalism*. Oakland: PM Press, 2016.
Natural Reality: Kunstleriche Positionen zwischen Natur und Kultur in Aachen. Stuttgart: DACO, 1999.
Nisbet, James. *Ecologies, Environments, and Energy Systems in the Art of the 1960s and 1970s*. Cambridge, MA, London: MIT Press, 2014.
Nye, Joseph S. Jr., "Public Diplomacy and Soft Power." in D.K Thussu (ed.), *International Communication: A Reader*, 333–344. New York: Routledge, 2010.
Orlando, Sophie. *Art et mondialisation. Anthologie de textes de 1950 à nos jours*. Paris: éditions du centre Pompidou, 2013.
Pavese, Cesare. *La spiaggia*. Turin: Einaudi, 1942.
Pelton, Joseph N. "Satellites as Worldwide Change Agents." in D.K Thussu (ed.), *International Communication: A Reader*, 13–35. New York: Routledge, 2010.
Perulli, Paolo. *Terra Mobile*. Turin: Einaudi, 2014.
Radical Nature: Art and Architecture for a Changing Planet: 1969–2009. Köln: *Walther König*, 2010.
Riley, Christopher. "Happy Birthday Earthrise." *BBC News*, December 24, 2008. retrieved on August 10, 2016 from http://news.bbc.co.uk/go/pr/fr/-/2/hi/science/nature/7797439.stm.
Robbins, Paul. *Political Ecology: A Critical Introduction*. Vol. 16. Hoboken: John Wiley & Sons, 2011.
Robertson, Roland. *Globalization: Social Theory and Global Culture*. New York: Sage, 2006.
Sallmann, Jean-Michel. *Le grand désenclavement du monde: 1200–1600*. Paris: Payot, 2011.
Sassen, Saskia. *A Sociology of Globalization*. New-York: Norton & Company, 2007.
Schliephake, Christopher. *Urban Ecologies: City Space, Material Agency, and Environmental Politics in Contemporary Culture*. Lanham: Lexington Books, 2014.
Schmid, Christian "Travelling Warrior and Complete Urbanization in Switzerland: Landscape as Lived Space." in Neil Brenner (ed.), *Implosions/Explosions: Towards a Study of Planetary Urbanization*, 90–108. Berlin: Jovis Verlag, 2014.
Scholey, Keith. "A History of Half a Century of Wildlife Television and Its Impact on Audiences." in Ian Convery and Peter Davis (eds.), *Changing Perceptions of Nature*. Martlesham: Boydell Press, 2016.
Serres, Michel. *Le Contrat naturel*. Paris: Flammarion, 1992.
Siewers, Alfred, K. *Re-Imagining Nature: Environmental Humanities and Ecosemiotics*. Plymouth: Bucknell University Press, 2013.
Sloterdijk, Peter. *Spheres II. Globes*, 783–818. South Pasadena: Semiotext, 2014.
Sobchack, Vivian. "Comprendre les écrans: une mediation *in medias res*." in Mauro Carbone, Anna Caterina Dalmasso and Jacopo Bodini (eds.), *Vivre par(mi) les écrans*, 29–46. Dijon: Les presses du Réel, 2016.
Steger, Manfred B. *The Rise of the Global Imaginary: Political Ideologies From the French Revolution to the Global War on Terror*. New York: Oxford University Press, 2008.
Stopford, Martin. *Maritime Economics*, Third edition, 35–43. London, New York: Routledge, 2009.
Swyngedouw, Erik. "Globalisation or 'Glocalisation'? Networks, Territories and Rescaling." *Cambridge Review of International Affairs* 17, no. 1 (2004): 25–48.
Swyngedouw, Erik. "Circulations and Metabolisms: (Hybrid) Natures and (Cyborg) Cities." *Science as Culture* 15, no. 2 (2006): 105–121.

Taylor, Charles. *Modern Social Imaginaries*. North Carolina: Duke University Press, 2004.
Tomlinson, John. *Globalization and Culture*. Chicago: University of Chicago Press, 1999.
Tomlinson, John. "Globalization and Cultural Identity." *The Global Transformations Reader* 2 (2003): 269–277.
Verne, Jules. *Le tour du monde en quatre-vingt jours*. Paris: Hetzel, 1873.
Wharton, W.J.L. (ed.). *Captain Cook's Journal During His First Voyage Round the World Made in H.M. Bark 'Endeavour', 1768–71*. London: E Stock, 1893.
Worster, Donald. *Nature's Economy: A History of Ecological Ideas*. Cambridge: Cambridge University Press, 1994.

Part I

Forgotten Spaces

1 Beyond Narcissus

The Sea and the Metamorphosis of Port Cities in the Late 20th Century

Gabriel N. Gee

In late July 2005, a *Ghost Ship* designed by the American artist Chris Burden made its way from Fair Isle, Scotland, to Newcastle-upon-Tyne, North-East England, unmanned, self-navigating via GPS.[1] In the mouth of the river Tyne, it sailed solitarily under the city's Millenium bridge, in what was once the busiest harbor for coal distribution in England. *Ghost Ship* brought to the fore the radical transformations of global maritime cultures and spaces that took place in the second half of the 20th century. With the introduction and subsequent adoption of the shipping container in the 1960s, the perception of maritime spaces evolved dramatically as container terminals replaced former docking areas, as automatization displaced harbor working communities and merchant navigation disappeared into a network of invisible motorways of the sea. Additionally, the 1960s witnessed a seminal shift in passenger transportation, with air travel overtaking maritime routes.[2] Jointly, these developments in the movement of goods and people induced a transformation of port cities' imaginaries. This chapter looks at a series of artistic practices which have engaged with this profound metamorphosis taking place at the turn of the 1960s–1970s, in order to approach and unpack our changing perceptions of maritime spaces. The aesthetic representation of the relation between port cities and the sea has a long history. In the modern age and in the Western art historical tradition, the emergence of a new pictorial genre in the 17th century, the *marine* or seascape, embodied a novel perception and recognition of nature.[3] The marine also significantly developed alongside the rise of the Dutch Golden age and its global commercial empire; hence it also articulated a world view with far reaching political consequences. While maritime painting is no longer a potent artistic genre, new forms of the marine have emerged in the late 20th century, which provide unique perspectives into the changing configurations of maritime socio-natures. To guide us into this reflection, I will revolve in particular around Allan Sekula and Noël Burch's documentary film *The Forgotten Space* (2010), which explores the impact of the container on global maritime economic and cultural spaces.

Additionally, I refer to the mythological figure of Narcissus, to explore the evolution of port cities' imaginaries. Narcissus, as the poet versed, fell in love with his own reflection mirrored in the liquid surface. Throughout history, Western port cities have emulated this mythological abandon in a reflective gaze. They and their adventurers were fascinated by the world beyond the seas; however, to what extent was their encounter with these worlds mediated by a projection of their own cultural expectations, fantasies and internal political ambitions? And to what extent did the caesura between sea and port in the 1960s invert this relation, as the sea was momentarily forgotten, while the world beyond channeled by emigration and cultural exchanges suddenly emerged at the heart of the port city? Port cities and coastal regions around the globe have been attracting increasing populations, and at a time of challenging global environmental issues, a multifaceted understanding of port cities' cultural relations to the sea and the worlds beyond appears crucial to assert a critical renegotiation of their territorial inscription in their natural sites and cultural histories. I will evoke first the figure of Narcissus, and its metaphorical value for an understanding of maritime imaginaries; secondly, I will consider the sea as a ghostly presence in the age of the container, before exploring two possible articulations of a *beyond* Narcissus: on the one hand a Western path described as an 'inverting' of civilization, and on the other a Southern hemisphere alternative in the form of the Taiwanese Lily.

1 Foreword: Narcissus and Port Cities

An analogy between the figure of Narcissus and the European port city serves as a starting point for this reflection. In Ovid's Metamorphoses, Narcissus falls in love with his own reflection in a pool, and only the words of the nymph echo his ultimately deadly fascination with his mirrored image.[4] This pool can be seen as the sea, a surface of projection not only for the sailing ships of the adventurous travelers, but also for the desires fomented by the European port cities as they launched themselves into the conquest of spices and foreign lands. The arrival of Spanish and Portuguese vessels in the New World and the Indian Ocean constituted a decisive step in the 'great opening up' of the world taking place at the turn of the 15th and 16th centuries.[5] The principles of an 'économie-monde' described by Fernand Braudel and originating in the Mediterranean Sea were expanding to new shores, thereby bringing about novel forms of global interconnectedness.[6] However, despite a leap into unknown spaces, the European gaze was also fueled by internal visions that limited the reach of any encounter with the Other. Sanjay Subrahmanyam evokes for instance the fanatic messianic vision of the King of Portugal Don Manuel, and the subsequent ever present quest for the Kingdom of Prester John that accompanied Vasco de Gama's first encounters on the Indian continent.[7] In understanding the modes of representation of past and present Western societies, pictorial production offers a privileged entry into the narcissist character of these great discoveries.

For the age of the great European discoveries was paralleled by the advent of a new system of representation in Western culture, based on the use of perspective in the construction of a picture plane operating as a window open onto the world. Interestingly, Leon Battista Alberti, the major theoretician of this new pictorial space, specifically referred to the figure of Narcissus in his treaty on painting, *Della Pittura* (1436).[8] Alberti sees in the mythological figure the inventor of painting, 'the flower of all the arts,' and concludes by asking: "what is painting but the act of embracing by the means of art the surface of the pool?" As commentators have underlined, Alberti's reference to Narcissus is allegorical; it aims to highlight the demiurgic power of painting.[9] The allegory itself also incarnates the manner in which the aesthetic system which emerged during the Renaissance accompanied the shaping of a modern Western cosmological conception in the age of discoveries. Through the use of linear perspective, the three-dimensional world finds itself projected onto a two-dimensional surface in a highly illusionistic manner, suggesting a truthful and objective representation. However, as Erwin Panofsky famously underlined, the system of perspective remains a *constructive* device, symbolic of the society which invented it, and its specific intellectual conception of the world.[10] Similarly, the set of relations that the European adventurers formed with the 'new world(s)' was informed by ambiguous purposes. The journal reporting the travels of Joris Van Speilbergen around the world, significantly entitled *The East and West Indian Mirror* (1619), aimed to present to the reader "as in a mirror some strange things that the art of navigation has brought to light," and to enlighten him in "foreign countries, peoples, nations and trade, just as if he had visited the same in person."[11] Voyages, such as those of Captain Cook in the 18th century, hosted a personnel devoted to documentation, of space, flora and fauna, including artists.[12] Yet as Geoff Quilley pointed out, while such members of the crew professed noble ambitions, their participation was also motivated by career opportunity and financial gain, just as the travel's aforementioned ethos of furthering knowledge was rooted in aggressive commercial competition.[13] Behind the noble claim to represent the world as it appeared to the bold adventurers, lurked a narcissist projection that tended to mediate the perception of foreigners and foreigness.

A pictorial genre emerged to bear witness to the expansion of European maritime networks: the *marine*. Margarita Russell, in her study of Hendrick Cornelisz Vroom, the so-called father of Dutch marine art, aptly retraced the direct filiation between the growing interest in naturalistic representations of the sea, evident in religious paintings of the 15th and 16th centuries, and the appearance of an independent maritime pictorial genre in the Low Countries.[14] Within the broader pictorial genre of *landscape*, which itself is closely associated to the Dutch Golden Age, the seascape rapidly came to represent a *Mirror of Empire*.[15] As the Dutch, soon followed by the British, sent their ships across the seas to take part into the lucrative Asian and American trades, the marine could articulate the quintessential relation that the two nascent maritime nations were establishing with the world at large. Marco Vegetti, rereading Carl Schmitt's *Land and Sea* (1942), points to the new world vision that underlaid the Western technological and industrial development, in which inland and ports have become a series of mere coastal connective interfaces within a deterritorialized network.[16] Christina Payne, in discussing subsequent 19th-century British depictions of the sea, could zoom into John Brett's *Britannia's Realm* (1880), to identify the painting's representation of a serene sea gently punctuated by a few sails on the horizon, as a quintessential symbol of the British empire.[17] Within this new spatial order, the port city despite its far reaching tentacles, secretes its understanding of the world from within. In *Skyline. La ville Narcisse*, Hubert Damish evokes the possibility of a modern urban Narcissus, and explores the modalities of this urban inward turn.[18] Our narcissist port city is not turned towards the sky, however, but towards the sea, and its mirroring trope should correspondingly be called 'sealine.' Within a history of globalization, the mirroring sealine became associated with the generic imposition of Western *formats*.[19] However, to revert to the Ovidian tale, as Alina-Daniela Marinescu reminds us, Narcissus ultimately realizes the illusion of his infatuation, that it is himself with whom he has fallen in love. The poet grants him the crucial capacity to become aware of himself, the precondition through which to become conscious of the world.[20] Similarly, the narcissus port city saw its spell broken at the turn of the 1960s.

2 Spectral Navigation and Invisible Harbors

In the opening shot of *The Forgotten Space*, Allan Sekula and Noël Burch's 2010 documentary film, the camera floats in the mouth of the Meuse, the much remodeled river by Rotterdam Harbor; cranes are visible in the distance, then two power plants towers, smoking, the sound of ships and seagulls. The narrator's voice enters: we are mid-stream, in "a muddy estuary near a port," a "forgotten space, out of sight, out of mind." This spectral maritime space is located in an in-between, with the hinterland upstream, "a greedy continent," restlessly demanding goods to be fed, and mirroring cities downstream, "other ports, great harbour cities," all connected through the oceans and a hidden economy, "one thousand invisible ships, one and a half million seafarers, binding the world through trade."[21] Considered in relation to the photographic oeuvre of Allan Sekula, *The Forgotten Space* condensed the Los Angeles photographer's reflection on maritime spaces identified as the locus of drastic and unequal globalization forces.[22] The sea has become invisible, a mere gliding surface on which ships move according to preordained routes between similar functional port-container terminals. Yet, there is a labor involved on board those ships, and awaiting their cargo: the sequence dedicated to Sekula's home town focuses mainly on the Hispanic community servicing the trucks that carry the containers inland to the consumers. Sekula in that respect rejects the assumption that with the advent of the age of information, manual labor just disappeared; the so-called 'post-industrial' society has on the one hand simply displaced industrial working sites, and on the other intensified the consumption drive inherited from the 19th century. It is this trans-industrial superposition characterized by uneven benefits that *The Forgotten Space* explores. Crucially, Sekula and Burch's investigation revolve around the impact on maritime transportation of the shipping container, which was rapidly adopted in the 1960s. Marc Levinson in his study on *The Box. How the Shipping Container Made the World Smaller and the World Economy Bigger*, reminds us of the poetical mechanics of the pre-container age, the world of the longshoremen, of hooks and scrambles, and of the high cost entailed by maritime merchant transportation.[23] With the worldwide adoption in the 1960s of the standardized container, maritime transportation became affordable and devilishly efficient.[24] It also transformed port cities, and their relation to the sea. The historical morphological mutations of port cities have been precisely described by geographers in the late 20th century, with the literal separation between the port and the port city constituting in Brian Hoyle's analytical scheme the fourth historical stage in harbor development.[25] This stage is characterized by the abandonment of the infrastructures inherited from the industrial revolution, supplanted by container terminals enabling the rapid transit of goods on and from ships to and from trucks and rails. In this process, the sea tended to recede from the port city. In one of its most characteristic metamorphoses, it has become an ornament in the form of the marina.

Chris Burden's *Ghost Ship* evoked at the outset of this reflection, embodied the *spectral navigation* characteristic of contemporary maritime transportation, guided by electronic automation rather than sextants and astrolabes, erasing human bodies from the maritime scene. It also captured the new *deserted* harbors: no boats are to be seen in the Tyne's mouth. Instead of a flurry of sails depicted by William Turner in an 1823 watercolor, only *promeneurs* are to be seen on the Newcastle and Gateshead's refurbished quays, crossing the river via a Millenium bridge to visit the Baltic Centre for contemporary art, formerly a silo building, or going to listen to a concert at the Sage, the nearby wave-shaped concert hall which opened in 2004. On the other hand, the newly established industrial infrastructures which have replaced the 19th-century sites present a different type of deserted scenery. The French artist Marie Reinert explored one such restricted areas in her video film *Quai* (dock), made in Fos-sur-mer near

Marseille (2012–2014). The film opens on a close up of a very large hull, but we move rapidly to a second sequence, a long tracking shot over petrol containers, a seemingly interminable infrastructure devoid of any human presence (Figure 1.1). This is the 'quay,' the former busy docklands of the port city, now a desert:

> We are in a relation to landscape with the pipelines. The questioning that I wished to further was that of the invisible. At certain moments we come close to the action, between the boats and the pipelines. The sound is very important here, the sound of petrol in the pipelines. We are in a relation to the machine and the body of merchandise. I also make use of the surveillance instruments of the universe in which I find myself, which gives a particular texture to the film.[26]

The Petrol and gas harbor is an open space, but whose access is severely controlled. The investigator brings the camera up the circular towers, where we can appreciate the lunar landscape, which is "an in-between, a no-man's land, where, as always, surreptitiously, a strategic and political situation is unfolding."[27] Humans do appear, but in divers' protective gear. The use of surveillance cameras emphasizes a sentiment of *outer space*, while the final shot is illuminated by one radiant light, guiding the humans on the ground as the lighthouse had once guided the sailors onto the sea.

Sekula and Burch also looked for those hidden spaces where economic accumulation takes place, serviced by invisible hands. In Rotterdam, the camera of *The Forgotten Space* filmed the lonely workers still (at that time) needed to manipulate the cranes within the highly automatized system of the container terminal. In an essay written to accompany the film release, Sekula and Burch significantly pointed out that most "educated people in the developed world believed that material goods travel as they do, by air."[28] For the boom of air transportation for passengers also occurred in the 1960s, with the introduction of jet airliners. Thus simultaneously the standardized

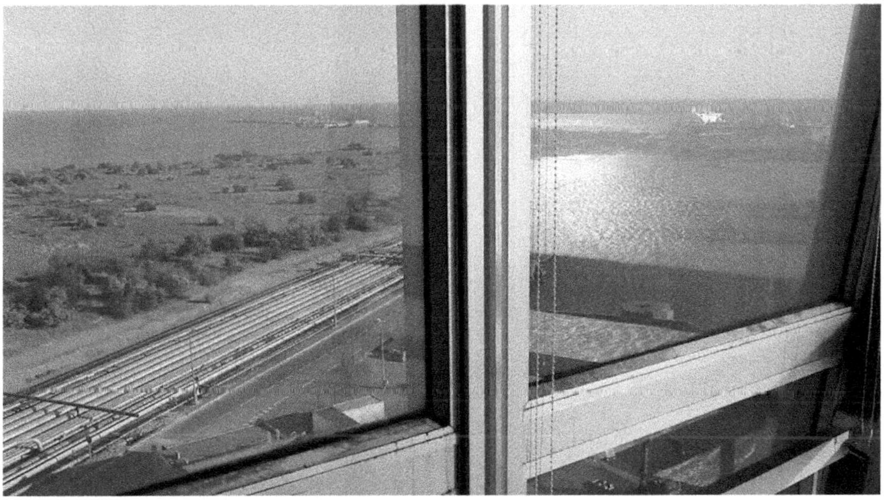

Figure 1.1 Marie Reinert, *Quai*, 2012–2014
courtesy of the artist

container displaced and to some extent suppressed the harbor of the harbor city, when air transportation was removing the sea as a significant conduit for the movement of human bodies in space. Sekula in his prolific work and reflection on maritime spaces made another crucial parallel between the hidden motorways of the sea, and the new virtual seas of information. In *Dear Bill Gates* (1999), Sekula writes a letter to the Microsoft founder, accounting for a swim past the entrepreneur's house. He affirms that he would have liked to see a recent purchase of Gates, a painting by Winslow Homer entitled *Lost on the Grand Banks* (1885), depicting two fishermen struggling on a frail boat in the Northern Atlantic. The end of the letter takes the form of a series of questions: why would Gates be interested in such a painting? Surely he realizes that these fishermen are going to die. Which prompts the subsequent analogy: "And as for you, Bill, when you're on the net, are you lost? Or found?" A question which also applies to us: "and the rest of us—lost or found—are we on it, or in it?"[29] The parallel is historical, for as the world lost sight of the sea, a forgotten space, it discovered a new ocean made of virtual digital connections. But it is also ontological, in that the mechanisms of both spaces are, for Sekula, governed by a problematic submission to economic imperatives, where the relevance of the 'social' is bluntly denied. In this world, the artist has to adopt the posture of an 'oceanic diver,' to better hear the social rumblings beneath the deceiving flatness of the computer screens.

It is in such an exploration of the crossroads between physical and virtual seas, that the Dublin based artist Cliona Harmey installed in 2015 two digital billboards in the center of the Irish capital, on which the name of ingoing and outgoing boats in the harbor appeared live (Figure 1.2). The intervention was made possible by a collaboration with the Port of Dublin Authority, granting the necessary access to its electronic transportation information. The billboards reintroduced into the town center

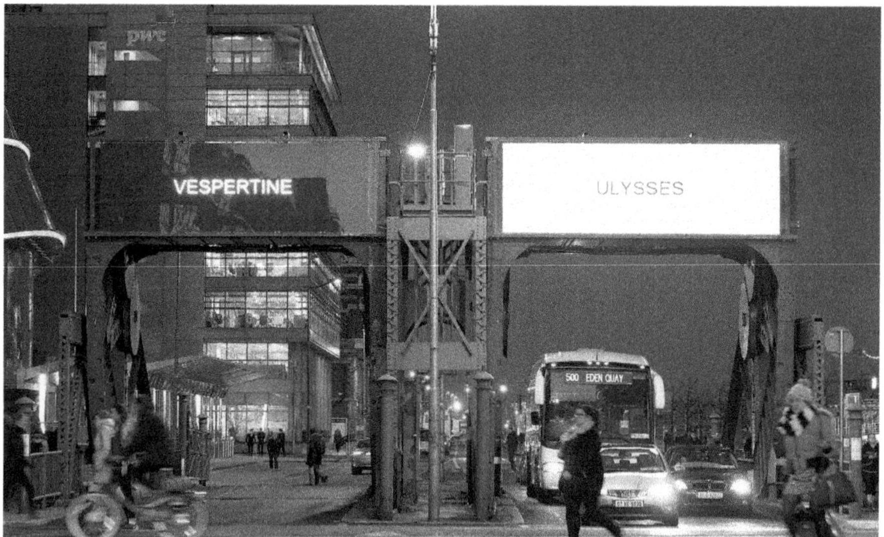

Figure 1.2 Cliona Harmey, *Dublin Ships*, 2015
courtesy of the artist

the invisible ships going in and out of a harbor now located at the end point of the river Liffey. They did so in what was once the heart of industrial maritime Dublin:

> 'Dublin Ships' has been designed so it could be seen by commuters over an extended period of time at a busy inner city intersection. There are six thousand cars that go down there a day, it's very busy, and it's part of the redeveloped docklands, the docklands were in a derelict state in the 1980s. Now it is in the heart of google and twitter, who have settled their headquarters close to that part of the river. The structure which supports the signage panels is an old Scherzer bridge. It was used to speed up water traffic, before it was installed it took something like half an hour to let a barge go through (quayside traffic was stopped), this bridge reduced that to a few minutes. Hence it is in a way linked to speed. It is obsolete now and not used anymore but is a heritage structure.[30]

The work stems from Harmey's longstanding interest in communication systems. Again we see here the conflagration of two technological developments, in which the physical modification of good transportation is paralleled by a revolution in channels of representation. We are reminded of the presence of the sea and our material lives sailing by, through digital ponds connected to digital oceans. If spectral navigation and ghost harbors were the initial result of this dual evolution, thereby to some extent halting the narcissist gaze of the port city in that it suddenly turned inwards, a subsequent opening from within could lead to a possible *beyond.*

3 'Inverse Civilizations': A Passage Beyond Narcissus

For if Narcissus, in the Ovidian tale, comes to realize the mistake of his infatuation, he drowns nevertheless. To prevent such a fate, port cities have to move beyond the mirror stage. Homi K. Bhabha in the introduction to his *Location of culture* (1994) situates our epoch's resolutely in the age of the *beyond* (emblematically captured by the proliferation of a post-terminology: postmodernism, postructuralism, post-condition . . .), characterized by a novel awareness of the subject positions enabling a shift away from the 'singularities of class or gender' and the 'narratives of originary.' It is in this shift towards a reflection on the processes at play in the production of cultural differences and the appropriation of in-between interstitial spaces that new emancipated sites of communal contestation and collaboration can be drawn.[31]

In May 1992, a fleet of two hundred ships coming from Genoa and Lisbon regrouped in Cadix, before sailing towards Porto Rico. The 'Grand Regatta Columbus' commemorated the five hundred years of the discovery of the Americas by the Genovese sailor Christopher Columbus. Having crossed the Atlantic, the tall ships visited Boston and New York, before sailing back to Europe, and reaching Liverpool. As a counterpoint to the event, the artist Keith Piper coordinated a series of interventions and displays that year in the port cities of Bristol, Hull and Liverpool, entitled *Trophies of Empire*. As Piper recalled in a text entitled *Shifting Landscapes*, *Trophies of Empire* aimed to provide a counter-narrative to the imperial narrative evident in the regatta, taking at its cue Margaret Thatcher's famous 'swamping speech' of 1978, in which the soon to be prime minister had declared that "the British character has done so much for democracy, for law, and done so much throughout the world that if there is any fear that it might be swamped, then people are going to react and be rather hostile to those coming in."[32] The speech alluded to the threat of an 'inverse colonization.' The various works commissioned for *Trophies of Empire* explored on the other hand counter-narratives of the imperial history and the complexities and textures of the British national fabric. In Liverpool and Bristol in particular, artists unearthed the somber past of the cities' busy involvement in the triangular trade, based on the exchange of goods, enslaved humans and sugar. Sculptor Juginder Lamba installed *The Cry* (1993) in the courtyard of the Bluecoat Gallery, a work comprised of a rectangular cell made of oak timber and metallic chains. Arms outstretched in despair emerged behind a locked door, a dark reminder of the slave trade at the heart of Liverpool's 18th-century economic fortune. On a more sarcastic note, the collective South Atlantic and Trouble designed a cabinet of curiosities displaying sugar, tea and cigarettes, as well as a souvenir set of picture cards to commemorate the "fantastic story of five hundred years of imperial violence, pillage and mayhem around the globe." In Bristol, an installation piece by Carole Drake entitled *Commemoration Day* revealed invisible historical attributes of one of the city's father figures, Edward Colston. Combining chrysanthemums—a favorite flower of Colston—left to wither on the floor, a screen projected photograph of the girls at Colston School covering his statue with flowers on Colston Avenue in 1973, as well as a souvenir statue of the political figure; the installation was accompanied by a soundtrack based on the school hymn 'Rejoice ye pure in heart,' which was progressively disrupted by an enumeration of slave deliveries to the West Indies, in which Colston had been directly involved as a merchant. *Trophies of Empire* was exemplary of an aesthetic articulation of maritime history capable of revealing the muted voices of 'the Black Atlantic,' to refer to Paul Gilroy's seminal identification of the transatlantic culture emerging from the triangular trade. To a fear of 'an inverse colonization,' it offered the reality of 'an inverse civilization.'

In the specific emblematic context of the Merseyside Harbor, the collective Visual Stress participated to *Trophies of Empire* with a piece entitled *Mobile Auto Mission*. Visual Stress had developed in the late 1980s a form of performative intervention dubbed an 'Urban Vimbuza,' after an ancient Zambian ritual. In 1988, for the *Urban Vimbuza—Death by Free Enterprise*, Visual Stress coordinated dancers, performers, drums and industrial proto-techno music as well as Welsh climbers on the Bluecoat Gallery site to perform a purification of the city. For Vimbuza means 'possessed,' and refers to a sickness of mind that characteristically affects local population after the arrival of colonial settlers. It is characterized by a mimicking of the behavior of the new masters. The alienation can be cast on condition that the patient realize their sickness, followed by a healing ritual, in which are used "Vimbuza songs to reach into the patient mind, drums to summon up the forces of healing, dance to cast out the spirit of possession." In the British port city however, at the end of the 1980s, the ritual targeted another yet not so different type of alienation:

> Switch venue, from village to a city, psychogeographies of sinister histories, in Liverpool in particular, it was easy to see the lines of slavery that stretched across the centuries, and the races. We may have our shopping trolley, our washing machines and microwaves, cars, (to get us to work) but at root we're all still slaves, we just swapped our chains for Pay As You Expire wage slavery.[33]

The sickness within the Western port city resided first in the traces of dark heritage, which unaddressed pursued their latent infectious spell, which we have characterized by a form of narcissist blindness. Secondly, the sickness has also mutated in a local form: the society of consumption where our addictions are fed by the proliferation of goods and a desire of earthly possession fueled by visual advertising. The so-called process of civilization is inversed, and it is the commodification of the European city and the alienated condition of its inhabitants that must be fought by shamanic incantations.

4 An Asian Lily: A Flower for Our Times

So far I have evoked the process through which port cities in the West in the second half of the 20th century became separated from their ontological relation to the sea. And I have considered how in this radical global transformation, the Western port cities might find a remedy to the self-infatuation that was largely blinding them to the plurality of the world across the oceans it sent ships over. This was made possible by a turn away from the disappearing mirroring gaze of the sea, to its own unacknowledged multiplicity: the port cities' 'home' (oikos) was not a monolithic haven impervious to time, but a textured palimpsest with multiple identities and allegiances. Immigrations, emigrations, transcultural exchanges and transformations had shaped a hybridity from within the port cities, that opened the door to a renegotiation of both their heritage and present identity. But what of port cities in non-Western worlds, how did the introduction of the standardized container affect their substance and their representations? Is the figure of Narcissus transferable? As a pure product of the Western cradle of civilization, Greek and Roman mythology, the obvious answer would seem to be: no. However, it is interesting to note that historians exploring the reaction and resistance of Asian civilizations to Western expansion, have often compared the two, coining the term 'the Asian Mediterranean.'[34] Modeled after Braudel's study on the Mediterranean as the birthplace of an eventually all-encompassing économie-monde, the much debated term suggests a parallel phenomenon taking place in Asia, often fragmented in various pockets of commercial networks.[35] In that respect, perhaps a figure such as Narcissus could help us understand port cities beyond European shores. Furthermore, the adoption of the box was a global phenomenon, impacting the form and imaginaries of port cities throughout the world correspondingly. In *The Forgotten Space*, Sekula and Burch eventually make their way to Hong Kong. At this crossroad of civilizations, they focus upon the Filipino workforce: men serving as crew on ships, women performing house care, as well as the Chinese croupiers, serving on floating casinos that sail without destination in the South China Sea. They then visit Shenzhen, on China's nearby shores. Once a city of three hundred thousand people, it has become an exemplary factory town of the Deng Xiaoping's era, characterized by economic reform and an opening to global fluxes. Shenzhen now houses some 20 million laborers, supporting an economy that supplies affordable IT products worldwide.[36] Production and consumption sites are interconnected now more than ever before. But are the maritime motorways equally as invisible on the production shores? And how do complex national and regional identities in those increasingly populated coastal regions negotiate the changing patterns of maritime circulation?[37]

Answering these questions would require a layered account of the multifaceted situations, identities and imaginaries of port cities as different as Hong Kong, Lagos and Osaka. Even so, I would like to briefly evoke one archetypal spatial crossroad of maritime exchange deeply tied to past and present global networks: Taiwan. To consider the changing representations of maritime spaces in Taiwan, I explore the work of Yang Mao-Lin, an artist whose practice has espoused the complex narrative of the island. The island of Taiwan is situated to the North-East of Hong Kong across a perilous strait. Since the 1950s, it has experienced an unprecedented economic surge, related to American political investment in the context of the Cold War (see also Tantinipankul in this volume).[38] Taiwan, however, had been at the junction of Asian and European navigational charts since the 17th century, and its identity, as that of Hong Kong, is irremediably linked to the maritime circulation and exchanges in Southeast Asia following the expansion of Western commerce and politics in the 17th century. Taiwan, as an island positioned at the crossroads of maritime routes, can be seen as a harbor where different socio-natures have become intertwined: indigenous, European, Chinese, Japanese, with their toll as well as their promises for human and non-humans alike. The Taiwanese artist Yang Mao-Lin has represented throughout his oeuvre the striking and multifaceted colonial history of the island, encapsulated in particular through his decisive 1990s series *Made in Taiwan*.

Made in Taiwan is composed of three sections, political, historical and cultural. In the first section, giants fight in an evocation of the Republic of China (ROC)'s struggled revocation of martial law in the 1980s.[39] The second section entitled *Zeelandia Memorandum* displays portraits of political, religious and military figures, who are associated with the colonization of the island by the Dutch, the Chinese and the Japanese from the 17th century onwards.[40] For instance the figure of Frederick Coyett (1615–1687) is depicted standing before a Dutch man of war and the coasts of Formosa, on which the menacing profile of the castle of Zeelandia can be seen (*Zeelandia Memorandum* L9302). The Zeelandia castle named after the low country province was the first fortress erected in Taiwan by the Dutch East India Company, and can still be seen and visited as a heritage site in Tainan in the south of the island (*Zeelandia Memorandum* L9302). But the portrait of the last Dutch governor of the island is also juxtaposed to that of Zheng Chenggong, also known in Europe as Koxinga, the heir of a powerful maritime family which retreated to the island in 1663 following their rebellion against the new Qing dynasty (ZM L9301). Similarly, the silhouette of a canon on which war ships seem to be sailing towards the island, is drawn together with the portrait of Li Hongzhang (1823–1901), the general and diplomat under whom Taiwan was ceded to Japanese rule following the treaty of Shimonoseki (1895) (ZM L9401). In the subsequent series of the historical section, *Yun Mountain Memorandum* (1991), *Lily Memorandum* (1993) and *Tayouan Memorandum* (1994–99), Yang explored the fauna and flora of the island:

> Two of the series inspired by animals and plants are "Yuanshan Memorandum" and "Lily Memorandum," which delve into the physical geography of Taiwan inhabited then by our earliest ancestors. Different colonizers came and went, and finally the new immigrants from the Mainland China arrived, animals and plants flourished and decayed along the tides of history. During this process, animals and plants of the land suffered the catastrophe together with humans. While people were exploited and killed, flora and fauna together with the land were also impoverished and destroyed. This should sound familiar to all the colonized around the world. That's why I chose Taiwanese endemic and native species, such as the Formosan clouded leopard, black bear, sika deer, and lily—the symbol of the aborigines. However, these symbolic meanings have been discarded

after the Han Chinese took over the island. And this is why they become motifs in my work.[41]

The botanic and animal series offer an allegorical reading of the multi-directional process of colonization and the offshoots of modernization. A close up on a shell, a symbol of early human settlements and civilization, is juxtaposed with an aerial view over the Taiwanese inland, crossed by rivers and punctuated by pre-colonization signs of habitation (*Yun Mountain Memorandum M9108*). The lily is also a recurrent motif in Yang's work, giving its name to a whole section of *Made in Taiwan*, in which the flowers are depicted in various perspective on large surfaces from close range *(Lily Memorandum L9401*) (Figure 1.3). In *Tayouan Memorandum*, the prehistorical past encounters the modern technological age: dinosaurs roam on the upper plane of *Tayouan Memorandum—Dragon* (L9501), while the lower part of the canvas is enframed within an orb, in which a submarine emerges next to a pagoda in a picturesque seascape. The combination of symbols incorporating Western and Eastern representational motifs makes for the creation of a *hybrid marine*. For besides the emblematic historical figures, Yang, crucially, depicts agents which have been essentially ignored by the forces of human history, including the aboriginal population of Taiwan. However, just as Taiwanese indigenous populations have survived the waves of colonization (notably in the Eastern mountainous part of the island), traces of the islands roots can be recognized and reincorporated in the contemporary fabric of the island

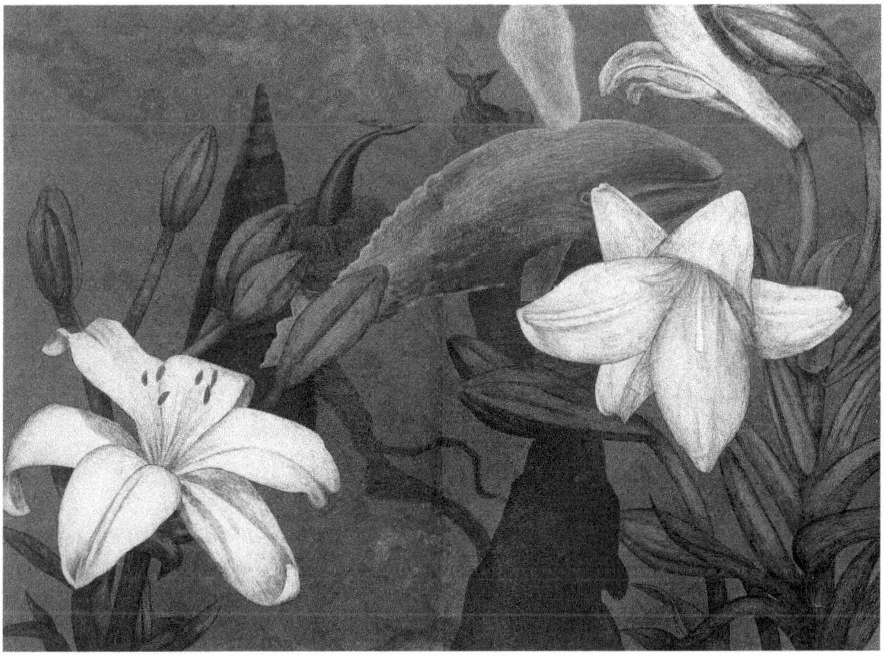

Figure 1.3 Yang Mao-Lin, *Lily Memorandum L9401*, 1994, oil and acrylic on canvas, 194 × 260 cm

courtesy of the artist and Tina Kang Gallery

in the age of computer chips 'made in Taiwan.'[42] The Taiwanese Lily, as a symbol of the survival and resurgence of the past into the present, embodying a state of hybridity that might suggest a move *beyond the beyond*, offers the possibility of abandoning the deathly Narcissus for a thriving future. It is to such alternate symbols, which are able to affirm the troubled and forgotten past of men and their environments, that an enlightened port city might turn its gaze to in the 21st century.

The introduction of the container in the 1960s induced a morphological metamorphosis of port cities, which was accompanied by a paradigmatic evolution of their relation to the sea and their urban and maritime imaginaries. The sea had long been for Western societies a space of 'terror and fascination,'[43] which the modern age progressively rationalized, leading to the controlling and exploitative stance of the 19th century characterized by extensive horizontal and vertical mapping and connectivity (the telegraph, submarine exploration, modern docks, port-terminals. . .). The radical ingenuity of the container infrastructure exacerbated this logic to the point that the sea became invisible: a transparent surface on which automated machinery carry the goods desired, produced and consumed by our societies, a forgotten space. Sekula and Burch reveal the underlying human toil behind this seemingly spectral veil, characteristic of the *trans-industrial* condition in which different industrial cultures are superposed and feed each other. The separation between port and urban core, however, opened the door to an emancipation of port cities from a form of narcissist blindness inherited from the age of the discoveries. This development which I described as a move beyond Narcissus, is to be seen in Western port cities realization of the difference *within*. To the monocular perspective of an 'inverse colonization' responds the positive reality of an 'inverse civilization.' This Western emancipation from itself, must be considered in parallel to alternate representations of global maritime interconnectedness. The work of Yang Mao-Lin offers one such multisided perspective on the encounter between different worldviews, as they came to shape the history and development of Taiwan from the 17th century onwards. In electing the Taiwanese indigenous lily as a symbol of a hybrid history encompassing resilient forms of survival from both humans and 'natural' species, Yang points to a possible negotiation of the complex textures of political and cultural formation. Through such symbols there also lies the possibility of rediscovering and reconfiguring maritime socio-natures in both an awareness of their complex interconnected histories, and a positive trans-generation that might prevent us from meeting the fate of the doomed Narcissus.

Notes

1 Chris Burden's *Ghost Ship* was commissioned by the Newcastle based art agency Locus +. <www.locusplus.org.uk/projects/68~Ghost+Ship/information>
2 Martin Stopford, *Maritime Economics* (New York: Routledge, 2009, first published 1997), 35–43.
3 Margarita Russell, *Visions of the Sea: Hendrick C. Vroom and the Origins of the Dutch Marine Painting* (Leiden: Brill, 1983).
4 Ovid, *Metamorphoses*, 8CE.
5 Jean-Michel Salmann, *Le grand désenclavement du monde, 1200–1600* (Paris: Payot & rivages, 2011).
6 Fernand Braudel, *La dynamique du capitalisme*, (Paris: Arthaud, 1985).
7 Sanjay Subrahmanyam, *The Career and Legend of Vasco da Gama* (Cambridge: Cambridge University Press, 1997).
8 Alberti, *Della Pittura* (1436).

9 Cristelle L. Baskins, "Echoing Narcissus in Alberti's *Della Pittura*," *Oxford Art Journal*, 16, 1 (1993), 25–33.
10 Erwin Panofsky, *Perspective as Symbolic Form* (New York: Zone Books, 1991).
11 Joris van Spielbergen, *The East and West Indian Mirror: Being an Account of Joris van Speibergen's Voyage Round the World (1614–1617) and the Australian Navigation of Jacob Le Maire*. Translated, With Notes and an Introduction by A.J.De Villiers, printed for the Hakluyt society (1906).
12 Geoff Quilley, *Empire to Nation. Art, History and the Visualization of Maritime Britain, 1768–1829* (New Haven, London: Yale University Press, 2011).
13 Figures of ambassadors that could bridge differences between the diverse cultures coming into contact certainly existed, such as explored by David Clarke in *Chinese Art and Its Encounter With the World* (Hong Kong: Hong Kong University Press, 2011), as well as interstitial figures such as those explored by Sanjay Subrahmanyam in *Three Ways to Be Alien: Travails and Encounters in the Early Modern World* (Waltham: Bradeis University Press—Historical society of Israel, 2011). The history of such interconnected interstitial figures and exchanges is of particular relevance to an understanding of the late 20th-century questioning of meta-narratives. Nevertheless, the reflection here focuses on the dominant framework within which the European great discoveries are understood to have taken place.
14 Margarita Russell, *Visions of the Sea*.
15 George S. Keyes, *Mirror of Empire. Dutch Marine Art of the Seventeenth Century* (Cambridge: Cambridge University Press and Minneapolis Institute of Arts, 1990). The revealing title of an exhibition of Dutch marine art of the 17th century held at the Minneapolis Institute of Art in 1990.
16 Marco Vegetti, "Terra/mare—Aria," in *Terra Mobile*, ed. Paolo Perulli (Turin: Einaudi, 2014), 53.
17 Christina Payne, *Where the Sea Meets the Land. Artists on the Coast in Nineteenth-Century Britain* (Bristol: Samson & Company, 2007), 32.
18 Hubert Damisch, *Skyline. La ville Narcisse*, (Paris: Seuil, 1996).
19 Jacinto Lageira, *L'art comme histoire. Un entrelacement de poétiques* (Paris: Mimesis, 2016), 243, "The discovery of the world by the West entailed its progressive shaping, ordering, subjection, and the fomentation of a text and a generic discourse to be imposed to other civilisations, and whose recent consequences are the present globalisation, which still functions (. . .) according to Western standards" (author's translation).
20 Alina-Daniela Marinescu, *Spécularité déformante. Sur les traces d'un paradigme anti-mimétique de l'art* (Paris: l'Harmattan, 2013), 46, with a reference to Christian Moraru, *Poetica reflectarii*; incercare in arheologia mimezei (Bucharest: Editura Univers, 1990).
21 Alan Sekulla and Noel Burch, *The Forgotten Space* (Doc.Eye film, 2010).
22 Jan Baetens and Hilde van Gelder, eds., *Critical Realism in Contemporary Art: Around Allan Sekula's Photography* (Louvain: Presses universitaires de Louvain, 2010, 2nd edition), 28–39.
23 Marc Levinson. *The Box: How the Shipping Container Made the World Smaller and the World Economy Bigger* (Princeton, NJ: Oxford, Princeton University Press, 2006).
24 The adoption of the container is in particular associated with the transportation entrepreneur Malcolm McLean, owner of Pan-Atlantic (then Sea-Land), and the company Matson on the West coast.
25 Brian Hoyle, "Global and Local Change on the Port-City Waterfront," *Geographical Review*, 90, no. 3 (July 2000), 395–417.
26 The author in conversation with Marie Reinert (7th of March 2016). "On se trouve dans une relation au paysage avec les pipelines. Le questionnement que je souhaitais prolonger était celui de l'invisible. A certains moments on s'approche de la mise en contact avec l'action, entre les bateaux et les pipelines. Le son est très important, ici le son du pétrole dans les pipelines. On est dans une relation à la machine et au corps de la marchandise. Je prends également appui sur les instruments de surveillance de l'univers dans lequel je suis, ce qui donne une texture particulière" (author's translation).
27 The author in conversation with Marie Reinert. Un paysage qui "est un entre-deux, un no man's land, où se joue comme à chaque fois, quelque part, mine de rien, une situation un peu stratégique, politique".

28 Allan Sekula and Nöel Burch, *Notes for a Film* (2010). <www.theforgottenspace.net/static/notes.html>
29 Alan Sekula, "Entre les mailles du net et la grande bleue (repenser le trafic des photos)," in *Titanic's Wake*, ed. Allan Sekula (Liège: Le point du jour, 2003), 11–37.
30 Cliona Harmey in discussion with the author (October 2015).
31 Homi K. Bhabba, *The Location of Culture* (London, New York: Routledge, 2012), 1–2.
32 Keith Piper, "Shifting Landscapes," in *Trophies of Empire* (Liverpool: Bluecoat Gallery, Liverpool John Moores University, 1994), 9–14.
33 Visual Stress, Vimbuzza, *The Ultimate Vaccination.*
34 François Gipouloux, *The Asian Mediterranean: Port Cities and Trading Networks in China, Japan, and Southeast Asia, 13th-21st Century* (Cheltenham: Edward Elgar, 2011).
35 Angela Schottenhammer (ed.), *The East Asian Mediterranean: Maritime Crossroads of Culture, Commerce, and Human Migration* (Wiesbaden: Harrassowitz, 2008); See in particular Wang Gungwu, "The China Seas: Becoming an Enlarged Mediterranean," 7–24.
36 Shenzhen, *From Factory of the World to World City* (Rotterdam: nai010, 2016).
37 John R. Gillis, *The Human Shore* (Chicago: The University of Chicago Press, 2012).
38 Lee Hsiao-Feng, *Histoire de Taiwan* (Paris: l'Harmattan, 2004), 115.
39 The Republic of China (ROC) took over Taiwan from Japan in 1945. In 1949, it migrated from mainland China to Taiwan, as it lost control to communists on the continent. Its leader, Chiang Kai-Chek, established an authoritarian regime.
40 *Yang Mao-Lin: Made in Taiwan, a Retrospective* (Taipei: Taipei Museum of Contemporary Arts, 2016).
41 Yang Mao-Lin in conversation with the author (8th of April 2016).
42 In the *Culture Section* of *Made in Taiwan* which began in 1996, Yang Mao-Lin adds an attention to television pop cultures, computer screens and connections, by representing and intersecting various figures from American and Japanese comics and animations. The *Culture Section* moves into an exploration of the 'virtual seas' which superpose themselves to the maritime words.
43 Alain Corbin and Hélène Richard (eds.), *La mer : terreur et fascination*, (Paris: Bibliothèque Nationale de France, Seuil, 2004).

References

Alberti, L.B. *Della Pittura.* 1436.

Baskins, Cristelle L. "Echoing Narcissus in Alberti's *Della Pittura.*" *Oxford Art Journal* 16, no. 1 (1993): 25–33.

Baetens, Jan and Hilde van Gelder (eds.). *Critical Realism in Contemporary Art. Around Allan Sekula's Photography*, Second edition. Louvain: Presses universitaires de Louvain, 2010.

Bhabba, Homi K. *The Location of Culture*, 1–2. London, New-York: Routledge, 2012 (first published 1994).

Braudel, Fernand. *La dynamique du capitalisme.* Paris: Arthaud, 1985.

Clarke, David. *Chinese Art and Its Encounter With the World.* Hong Kong: Hong Kong University Press, 2011.

Corbin, Alain, and Hélène Richard. (eds.). *La mer : terreur et fascination.* Paris: Bibliothèque nationale de France, Seuil, 2004.

Damisch, Hubert. *Skyline. La ville Narcisse.* Paris: Seuil, 1996.

Fairbank, John K. *China: A New History.* Cambridge, MA, London: Belknap Press of Harvard University, 1998.

Gillis, John R. *The Human Shore.* Chicago: The University of Chicago Press, 2012.

Gipouloux, François. *The Asian Mediterranean: Port Cities and Trading Networks in China, Japan, and Southeast Asia, 13th-21st Century.* Cheltenham: Edward Elgar, 2011.

Gungwu, Wang. "The China Seas: Becoming an Enlarged Mediterranean." in Schottenhammer, Angela (ed.), *The East Asian Mediterranean: Maritime Crossroads of Culture, Commerce, and Human Migration*, 7–24. Wiesbaden: Harrassowitz, 2008.

Hoyle, Brian. "Global and Local Change on the Port-City Waterfront." *Geographical Review* 90, no. 3 (July 2000): 395–417.

Hsiao-Feng, Lee. *Histoire de Taiwan*. Paris: l'Harmattan, 2004.

Keyes, George S. *Mirror of Empire: Dutch Marine Art of the Seventeenth Century*. Cambridge: Cambridge University Press and Minneapolis Institute of Arts, 1990.

Lageira, Jacinto. *L'art comme histoire. Un entrelacement de poétiques*. Paris: Mimesis, 2016.

Levinson, Marc. *The Box: How the Shipping Container Made the World Smaller and the World Economy Bigger*. Princeton, NJ: Oxford, Princeton University Press, 2006.

Marinescu, Alina-Daniela. *Spécularité déformante. Sur les traces d'un paradigme anti-mimétique de l'art*. Paris: l'Harmattan, 2013.

Ovid. *Metamorphoses* Garth, Dryden et al. (trans.). New York, London: Garland, 1976.

Panofsky, Erwin. *Perspective as Symbolic Form*. New-York: Zone Books, 1991.

Payne, Christina. *Where the Sea Meets the Land. Artists on the Coast in Nineteenth-Century Britain*. Bristol: Samson & Company, 2007.

Piper, Keith. "Shifting Landscapes." in *Trophies of Empire*, 9–14. Liverpool: Bluecoat Gallery, Liverpool John Moores University, 1994.

Quilley, Geoff. *Empire to Nation: Art, History and the Visualization of Maritime Britain, 1768–1829*. New Haven, CT, London: Yale University Press, 2011.

Russell, Margarita. *Visions of the Sea: Hendrick C. Vroom and the Origins of the Dutch Marine Painting*. Leiden: Brill, 1983.

Salmann, Jean-Michel. *Le grand désenclavement du monde, 1200–1600*. Paris: Payot & rivages, 2011.

Sekula, Alan. "Entre les mailles du net et la grande bleue (repenser le trafic des photos)." in Sekula, Allan (ed.), *Titanic's Wake*, 11–37. Liège: Le point du jour, 2003.

Sekula, Allan, and Nöel Burch. *Notes for a Film*, 2010, retrieved from www.theforgottenspace.net/static/notes.html.

Schottenhammer, Angela (ed.). *The East Asian Mediterranean: Maritime Crossroads of Culture, Commerce, and Human Migration*. Wiesbaden: Harrassowitz, 2008.

Shenzhen. *From Factory of the World to World City*. Rotterdam: nai010, 2016.

Subrahmanyam, Sanjay. *Three Ways to Be Alien: Travails and Encounters in the Early Modern World*. Waltham: Bradeis University Press—Historical soeicty of Israel, 2011.

Subrahmanyam, Sanjay. *The Career and Legend of Vasco da Gama*. Cambridge: Cambridge University Press, 1997.

Stopford, Martin. *Maritime Economics*. New York: Routledge, 2009 (first published 1997).

Van Spielbergen, Joris. *The East and West Indian Mirror: Being an Account of Joris van Speibergen's Voyage Round the World (1614–1617) and the Australian Navigation of Jacob Le Maire*. Translated, with notes and an introduction by A.J.De Villiers, printed for the Hakluyt society, 1906.

Vegetti, Marco, "Terra/mare—Aria." in Paolo Perulli (ed.), *Terra Mobile*, 49–78. Turin: Einaudi, 2014.

Yang Mao-Lin: A Retrospective. Taipei: Taipei Museum of Contemporary Arts, 2016.

2 Bangkok Against Its Nature

Worrasit Tantinipankul

The Bangkok metropolis is situated in lower central Thailand—a region long known for its abundant water networks producing rich soil, favorable to the cultivation of fruit orchards and rice fields, as well as mangrove forests along the coastline and providing habitats for local fisheries. However, the region is subjected to regular flooding due to the monsoon climate and heavy annual rainstorms. Furthermore, the downstream valley area of the nation's principal river, Chao Phraya, also faces rising seawater and inundation during the winter season causing alkaline soil and temporary shortages of fresh water. Although its natural condition is challenging for permanent inhabitants (human habitation), Bangkok has been a strategic area of the Thai state authority from the 16th century to modern times. Based on the control of water channels streaming down to the Gulf of Thailand, the strategic success of the early Bangkok settlement created a series of canal networks throughout the lower river valley region.

Bangkok remained a relatively small city until its development in the 1970s, which is the turning point for Bangkok's present expansion against its natural environment. This chapter reveals the changing conceptions of nature and urban settlement in Bangkok and its surroundings in the post-WWII era, during which the landscape of the city was significantly altered by rapid industrialization and modern development under American influence in the context of the Cold War. It also explores how different settlement patterns, from indigenous cultivation to modernization and industrialization, have transformed the landscape of Bangkok and its relationship to nature. Finally, it analyzes how these transformations have affected urban residents through severe environmental degradation and how historic communities have been rethinking their relationship with nature.

Bangkok: A Background

Before the 10th century, it is widely agreed among scholars of Southeast Asia that the lower region of the Chao Phraya River was under the sea. It gradually became a swamp plateau with rich soil accumulating at the delta of the River. The lower Chao Phraya River Valley became important in the mid-15th century when the Ayutthaya Kingdom was founded on the island at the confluence of the Chao Phraya River and the other two minor streams. The area along the river downstream from Ayutthaya to the Gulf of Thailand thus became a strategic area for participating in the lucrative sea trade passing through Southeast Asia. Several shortcuts were excavated to reduce the travel time, which resulted in the creation of networks of water channels that include the old oxbow loop of Chao Phraya and minor canals throughout the river basin. Villages and orchards had grown along the edge of the river and canals. Bangkok was one of the villages that got its name from its lavish orchards of water olives, and grew steadily as a check point for collecting taxes from the sea trade. It was registered as Thonburi in Ayutthaya's list of principalities. European mercenaries were assigned to build the fortress and city walls, while French troops were chartered to govern the town. Thonburi became the capital of the new kingdom of the Thais after the fall of Ayutthaya in 1776. The palace was built on the west bank of Chao Phraya River and by 1782, the palace moved to the east side of the river due to the dynastic change as Bangkok was proclaimed the new capital. From the 16th to 18th century, Bangkok developed through its advantageous use of water courses for trading activities. As it became the new kingdom's center with an increasing flow of internal and international trade, Bangkok began to exploit nature and its immediate environment even further. The two major rivers in the western and the eastern regions of the Chao Phraya River Basin were connected to Bangkok through extensive canal networks, opening up the land for sugar plantations and rice fields to feed the world's demand for sugar and other agricultural products. Bangkok became the center to which were sent the natural resources extracted in northern and central Thailand. With increasing colonial influences, the indigenous monarch complied with the colonial demands for abolishing monopoly trade, and favoured European models for restructuring the kingdom with direct control at the center. From the latter half of 19th century to the beginning of 20th century, Bangkok became more modernized adding colonial features such as baroque buildings, neoclassic shop-houses, bridges and boulevards to its urban landscape; but the canal network and orchards were still well preserved. The period between the 1920s and WWII saw some physical changes as the city wall and gates were demolished paving the way for new roads. Floating houses were prohibited on canals and riverfronts since their natural roof material was prone to catching fire.[1]

Bangkok as Bastion of the Free World in the 1960s and 1970s

Though the political changes and government policies in the 1950s initiated Bangkok's transformation, it was in the 1970s that the physical transformation of urban Bangkok against nature began. The absolute monarchy ended in a bloodless coup in 1932 which changed the country into a constitutional system.[2] However, political instability continued. From 1938 to 1944, there were few changes in the urban pattern of Bangkok, even though the population of Bangkok had grown faster than the nation's population rate.[3] Moreover, Thailand and its capital escaped serious damage during WWII. Though Thailand entered WWII on the Japanese side, it eluded severe war reparations by the US intervention. Unlike other major cities in Southeast Asia, China and Japan, Bangkok was not damaged critically by attacks from the foreign armies; it only incurred mild destruction of its only power plant and to the bridge crossing the river by bombs from allied warplanes at the end of the war.[4] As such, Bangkok from the mid-19th century to the war era was not altered by foreign powers in similar process to other neighboring colonial cities, but rather by its own ruling elites and their internal politics.[5] These were to bear greatly on the exploitation of nature and the strategies of urban development.

As an ally of the US and a bastion of the 'Free World' pitted against the rise of Communism in Southeast Asia, Thailand adopted development discourses steering the country into industrialization. Thailand's political and economic stability were in the

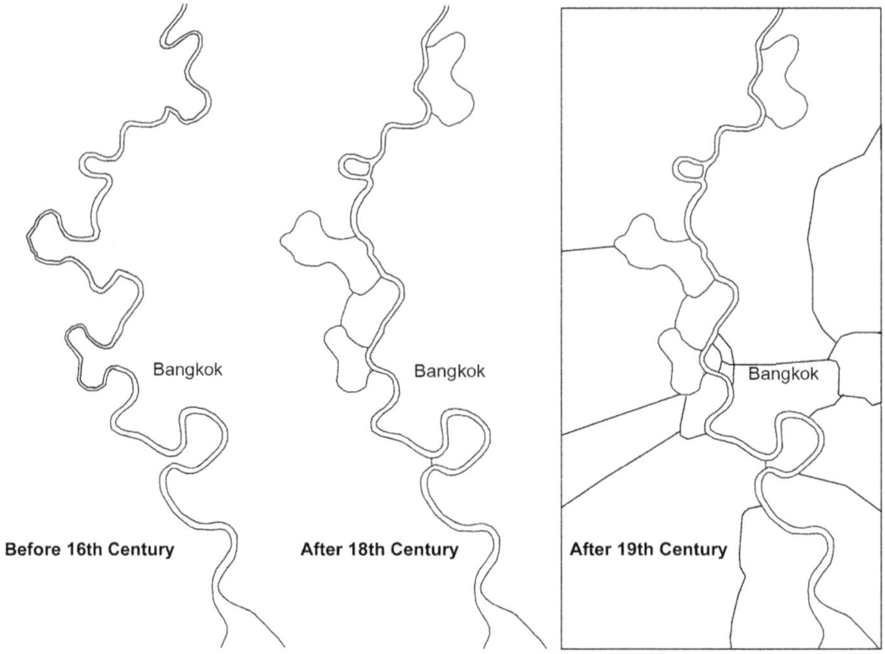

Figure 2.1 Worrasit Tantinipankul, Map Drawing of Lower Chao Phraya River and Bangkok From 16th Century to 18th Century, based on Steve Van Beek, *The Chao Prya: River in Transition*, (Kuala Lumpur: Oxford University Press, 1995)

US's interest as the latter engaged in the Vietnam War.[6] Thai military dictators were heavily supported from the 1950s to the mid-1970s further stimulating Thailand's economy.[7]

US military expenditure in Thailand also stimulated the service sector in Bangkok, especially in the tourism industry. From 1965 to 1972, US military personals spent an estimated US$ 111 million in Thailand for rest and recreation leave (R&R) which reduced significantly after the end of the Vietnam War.[8] From the mid-1970s, Japanese and East Asian investors and tourists filled in the vacancies in American tourism. Military leaders had prepared for a modern tourist economy in the late 1950s by establishing the Tourism Organization of Thailand, and the first national airline was created in 1959 and began flight service in 1960.[9]

The tensions during the Cold War Era made Bangkok and the central plain in Thailand a safe place in mainland Southeast Asia for tourism, and the industry continued to grow throughout the 1970s and 1980s, generating significant revenue and replacing agricultural exports. The nation's tourism agency was upgraded into the Tourism Authority of Thailand in 1979 to promote Thailand on a global scale.[10] The independent operation of the Civilian Airport became the Airport of Thailand in the same year. The highway connecting the airport to downtown was constructed in 1966 and the first international terminal completed in 1987.[11] From the 1980s to the 1990s, Bangkok's international airport at Don Mueang became the gateway for tourists visiting Thailand and became the busiest air transportation hub in Southeast Asia, second only to Singapore.[12] The expansion of the international airport contributed to the flourishing of housing developments in north Bangkok, which had once been a large green area connecting the city to the largest rice farming plateau of central Thailand.

American army tourism was the first to spark Bangkok's urban nightlife; resulting in the development of 5-star hotels and entertainment venues in the downtown business district, where the main arteries of Bangkok toward the east and southeastern regions are located. The influx of tourism turned the once quiet eastern suburb of Bangkok of Silom and Sathorn Road into a bustling entertainment, hotel and business district. The number of hotel rooms in Bangkok jumped four times between 1964 and 1970.[13]

A series of government-sponsored economic apparatuses were created to respond to the flow of US capital and investment. For instance, the nation's Board of Investment (BOI) was created in 1959 to facilitate the establishment of both domestic and foreigner's industrial development.[14] The National Economic Development Board (NEDB), which is the central agency responsible for devising Thailand's economic development strategies, was founded in 1958. Soon after the US supported Field Marshall Sarit Thanarat staged the coup and dissolved the parliament in 1958, the NEDB was upgraded (1959) and the first national development plan was launched in 1961.[15] This policy-planning agency became the major force in restructuring Thailand's economy in the 1960s and 1970s from an agriculturally based to an industrially based economy.

Between 1950 and 1975, the International Bank for Reconstruction and Development (IBRD), a branch of World Bank, provided US$ 440 million to Thailand for constructing highways, hydroelectric power and education.[16] The injection of capital significantly altered not only Bangkok but its natural environment. For instance, with the financial support of the World Bank in the 1970s, Bangkok launched the Greater Chao Phraya project, a large-scale development scheme aiming to provide irrigation through a series of reservoirs. The project was initiated in 1880 but wasn't realized until 1957. The first dam of the scheme, Chao Phraya, was finished by the loan from the World Bank.[17] The second reservoir, Yanhee, was finished in 1962 and the last dam, Sirikit, in the Nan Province was completed in 1972.[18] The project provided water to rice fields and helped control floods in the central plateau with Bangkok at its heart. Moreover, the dams also generated hydroelectric power for the Bangkok metropolis.

The flow of loans from the IBRD followed with the rise of direct foreign investment to Thailand. Under the support from BOI, about 1.25 billion baht foreign investment flew to Thailand annually between 1965 and 1975. Foreign firms invested a total of 6 billion baht with BOI-support program in the period of 1960–1982. Investments spurred the urban development of Bangkok in the eastern and southeastern suburbs of Bangkok, which became home to new factories. The Port Authority of Thailand was established in 1951 in the southeastern district of Bangkok and was to manage the growing import and export traffic into the new millennium.[19] New housing estates sprang up along the major arteries to the southeast, while slums emerged around the factories and the port of Bangkok.

In the late 1960s, direct foreign investment averaged 1 billion baht annually and it increased to approximately 1.5 billion baht in 1970s. The US was the leader in investing in Thailand during 1950s–1960s, but its investment declined afterwards and Japan became the major force of investment in Thailand in the 1970s and 1980s.

The 1970s saw the textile and automobile industry grow from Japanese investment. Thailand entered the 1970s with textile production increasing by six times. The automobile manufacturing industry has expanded steadily since the Thai government imposed the use of domestic parts for assembling every new car in the country. From 1960 to 1975, jobs in the automobile industry boosted from approximately 600,000 to 1.6 million and the number of medium size factories increased rapidly. Within a period of six years, the factories with more than 50 workers increased six times from 100,000 to about 600,000. This expansion also continued in the 1980s.

The mid-1970s was the crucial period in the economic development of Thailand and urban development of Bangkok. While the first half of the 1970s witnessed the continuing growth in value of agricultural export products, it dropped worldwide during the second half. The Thai government and business sectors closely collaborated to shift the industries from their reliance on import substitutes to exportation, thereby furthering industrial growth in the 1980s and 1990s. The majority of these new factory plants were established around the outskirts of Bangkok and adjacent provinces. The regional GDP in 1972 indicated that the Bangkok Metropolitan and its neighboring provincial area took 60% of the GDP in the nation's manufacturing sector.[20]

Bangkok during the 1970s expanded greatly toward the East in response to the booming investment in the manufacturing sector, modeled on the American automobile-oriented cities and interstate highways. Land-based transportation was chosen to become the backbone of the country. Urbanization rapidly expanded along the road networks throughout the central river valley region. These new developments, which had begun in the Cold War era, were influenced by principles of Western industrial urbanization, and amalgamated within the traditional aquatic urban space along the coastal suburban areas of the Gulf of Thailand.

Bangkok faced multiple urban challenges at the time. With the influx of labor immigrants searching for jobs in the booming manufacturing and service sectors, the city encountered serious housing shortage as slum-like communities sprang up around factories in the eastern and southeastern suburbs, as well as around major nodes of public facilities such as railroad junctions and the Bangkok Port. In 1971, it was estimated that more than 25,000 people lived in the area of the Bangkok Port at Khlong Toey. The population of Bangkok in 1970 had grown from 2.6 million in 1960 to 4.6 million and increased rapidly to 5.9 million in 1975. By the mid-1970s Bangkok's urban population had already doubled from the previous decade requiring significant increase in housing units. The built-up area of Bangkok also expanded greatly, from 43 square kilometers in 1936 to 184 in 1971.[21]

The government responded with the creation of the National Housing Authority (NHA) in 1973, providing housing for the poor and lower middle-income groups. In the early 1970s, slum communities were relocated from downtown to flats in the suburbs. But in late-1970s NHA turned its focus on the improvement of slum communities to be equipped with standard public facilities and utilities. By the mid-1970s, the private housing developers could still produce only half of the housing demand in Bangkok, although the number of housing developments jumped four times from 40 projects in 1974 to 154 in 1975.[22]

Moreover, most private-run housing projects targeted upper-income populations. Thus only 16% of Bangkok dwellers could afford them. These housing estates spread to the suburban area as a consequence of the rapid expansion of industrial development and service sectors. The traditional wooden stilt house to be found within

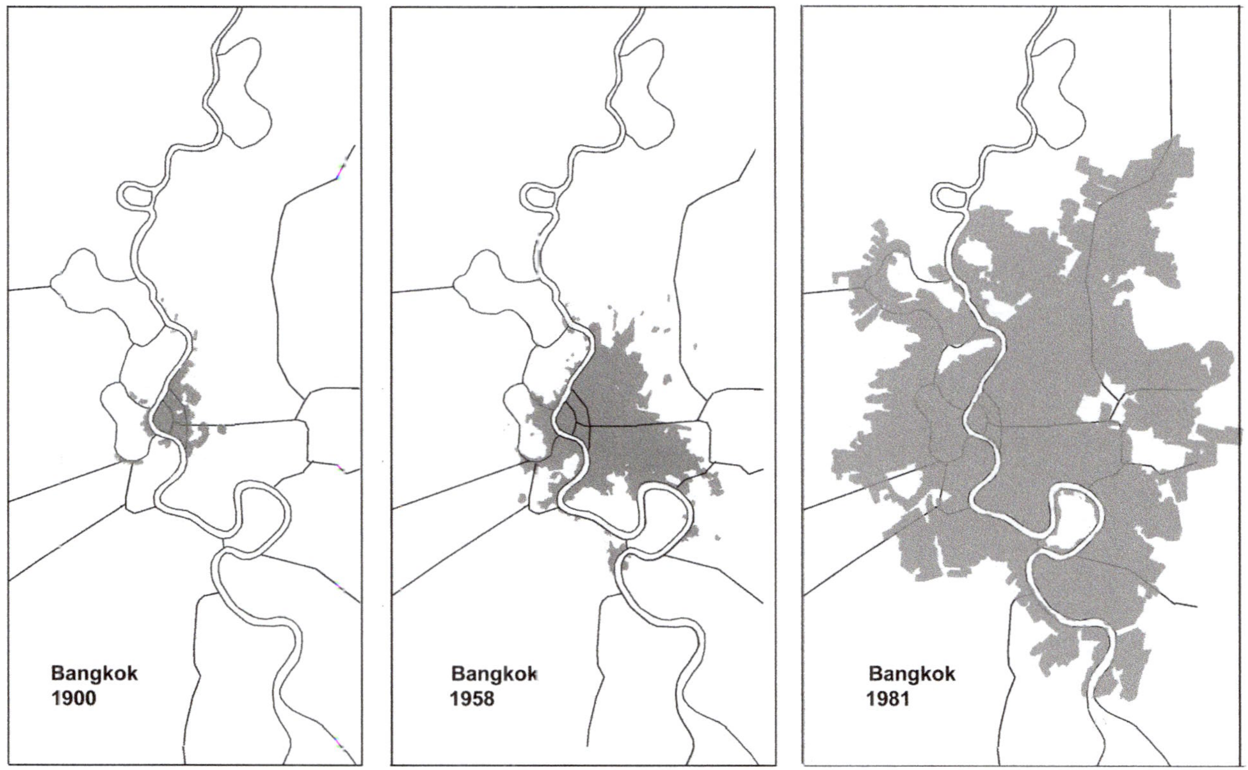

Figure 2.2 Worrasit Tantinipankul, Map Drawing of Bangkok Urban Expansion From 1900 to 1981, based on Larry Sternstein, *Portrait of Bangkok: Essays in Honour of the Bicentennial of the Capital of Thailand*, (Bangkok: Bangkok Metropolitan Administration, 1982), and Marc Askew, *Bangkok Place, Practice and Representation*, (London: Routledge, 2002)

farmland and along canals at the edge of the city disappeared. The housing estates had developed along the main arteries of the northern suburbs where the airport is located, the east, where new factories clustered, and the southeast where the major port at Khlong Toey was established. The majority of the luxury residential developments became gated communities with private security guards and high wall wrapping the estates, while upper middle-income housings transferred the ownership of facilities such as roads, electricity, tap water and sewage lines to BMA, resulting in an even more complicated road network.

At the end of the 1970s, the legislation of 1979 allowing a new type of high-density residence, the condominium, was passed and this high-density type of accommodation first emerged in Bangkok inner city to serve mostly expatriate communities.[23] This is the starting point for a new kind of urban dwelling space. Urban living quarters in Bangkok were once dominated by neoclassic shop-houses associated with Chinese and western communities in the historic district; Chinatown was now transferred to a condominium in the eastern districts of Silom, Sathon and Sukhumvit.

The 1970s also witnessed a commercial business and retail pattern that changed Bangkok to a mega-scale city. Shopping malls and complexes featuring hotels and other business functions emerged in the downtown area where European and Japanese expatriates lived. The modern mall patterns with retail facilities gradually replaced blocks of open-air market and shop-houses. This commercial development expanded into the outer suburb of Bangkok in the 1990s.

From 1971 to 1981, Bangkok's built-up area increased from 184 to 330 square kilometers[24] and by 1989, a built-up area of 1,568 square kilometers was recognized.[25] Bangkok from the 1950s to the 1970s had expanded without a clear direction. The urban aquatic life pattern was eradicated on the area east side of Chao Phraya River as the traditional canal system was either filled up, or replaced by sewage tunnels and more roads were constructed without well-defined planning, despite the production of a series of master plans advocating for the implementation of a drainage and sewage system.

A Greater Bangkok Plan B.E. 2533 (1990) had been devised in 1960 by the Boston based consulting firm, Litchfield, aiming to cope with a 4.5 million population and 780 square kilometers of urban area. Bangkok Metropolitan Authority's city planning division had also created a Greater Bangkok Plan B.E. 2543 (2000) in 1969 to cope with a population of 6.5 million and a built-up area of 935 square kilometers. However, without a land tax system and a functional real estate property assessment system linking land-use planning to law enforcement, the road network and public facilities were constructed following the expansion of industrial developments and housing estates led by interest of private sectors, ignoring any of these zoning regulations and land use plans. As a result, several parallel main arteries in Bangkok in the north, east and southeast zones were not connected together causing traffic jams at the junctions of each street linking the suburban residential estates to the main arteries of the business district and the downtown inner city.

With unplanned development, not only did a serious traffic problem emerge, but also an issue concerning the management of waste and storm water. Most storm water and wastewater from household units were drained directly to the same sewage line along small alleys connected to main tunnels under the main roads, which caused flash floods on the main streets during heavy rainstorm. With an expanding urban area, the Public Health Ministry claimed that only 2% of waste from households was treated properly while the rest was discharged into the canal system.[26] BMA also recognized that from 1990–2010, 75% of wastewater derived from domestic uses of residences, hotels, restaurants and markets.[27] In 1990, the Public Health Ministry found that communities in Bangkok disposed 180,000 kgs of solid waste and sewage water to canals and river every day.[28]

Bangkok also sank 1.70 meters down from 1920 to 1990 due to excessive use of underground water. When public water works were introduced in 1914, it served only 1,200 customers. However, Metropolitan Water work Authority (MWA) began to pump daily around 8,000 cubic meters underground water to distribute to users in 1954, and by 1982, the use had increased 55 times.[29] The growing housing developments, hotels and factories also consumed an additional 944,000 cubic meters a day.[30] The use of underground water was banned in 1977 and MWA announced a stop to underground water pumping in 1987. Nevertheless, the usage of underground water continued and aggravated the existing problem of flooding and water pollution. Bangkok's eastern suburb was below sea level and overflowing water had to be pumped to the Chao Phraya River while a series of water gates were installed in Thonburi, in western Bangkok, at the junctions of the canals and the river to protect the canals from saline water.

To aggravate the flood and water pollution, the major dams of the Greater Chao Phraya Project could not prevent overflooding from the northern main stream. At the beginning, the dams did succeed in doing so, but with the high degree of deforestation in the north due to illegal logging, the sediment increased in the dams faster than firstly expected shortening their capacity to hold water down to the river. Additionally, the dams could not stop flooding caused by heavy rainfall. Central plain Thailand and Bangkok suffered severe flooding in 1975, 1983, 1986, 1990 and 2011. Annual flooding is common for Bangkok residents after a heavy rainfall. Main roads in the city business districts then find themselves underwater, causing heavy traffic congestion.

The natural environment of Bangkok in the lowland of a river delta already subjected the city to seasonal flooding from the river and to rising saline water from the Gulf of Thailand. In order to keep urban economic activities prosperous, Bangkok was in need of a water retention system restraining storm water and preventing flooding disasters. A storm water retention system is crucial for any new real estate development that replaces the natural ground with impervious surface, in order to prevent an infiltration of the underground aquifer and an overloading of the urban drainage system. However, the retention system in Bangkok has never been imposed by building regulations. The Greater Bangkok Plan 2533 B.E. in 1960 preserved the green area at the eastern and western frontier of Bangkok for a retention basin and an existing canal system for the drainage of storm water. However, the master plan was never realized. The preserved green area which was supposed to be an urban park or farmland with a water retention system was eventually transformed into a new housing and commercial development.

Since floating houses were prohibited and cars replaced boats as the main transportation, canal networks were left unmaintained and stagnant with refuse from domestic uses. Canals in the past were the way for the population to connect to nature and a symbol of King Chulalongkorn's prosperity. But in this period, it came to symbolize disease and backwardness, helping to bolster arguments that the canals should be covered. Nature became a symbol of fear in the Cold War era. In addition, with the communist takeover in China in 1949 and other former Indo China nations, a fear of communist invasion through guerrilla warfare became rampant. The forest became a hiding place for communist insurgency during this era. In contrast to these wild and dangerous areas, road networks were seen as bringing technological progress, public services, trade and security.

Bangkok in 1980s: Subdue and Multiply

Bangkok in the first half of the 1980s experienced an economic downturn as the revenue from agricultural exports declined and export development was limited due to the oil crisis. The Thai government responded by promoting tourism, sending labor abroad and devaluating the Thai baht. Thailand was on the brink of bankruptcy as it had been supported with loans from the World Bank since the late 1970s. However, revenue from tourism grew from 3 billion baht in the mid-1970s to 18 billion baht in 1980 and, in 1985, reached 30 billion.[31] The turning point to the golden age of Thai exports came after a meeting at the Plaza Hotel, New York City, in September 1985. The five nations of France, West Germany, USA, UK and Japan discussed the destabilizing effect of the yen–US dollar currency exchange rate, as Japan enjoyed massive surpluses worldwide while the USA was deep in a deficit. If the situation continued, the world financial system could turn to chaos. The Plaza Agreement led to the depreciation of the dollar and the strengthening of the yen. To keep their products competitive, Japanese manufacturers relocated their factories from Japan to Thailand sparking an economic boom and a quadrupling of export value at the end of the decade. Synchronously, the slow growth in agricultural exports pushed 2 million rural laborers from the village to city, with half of them hired in temporary jobs in the expanding tourism industry and the textile and garments export industries.

The latter half of the 1980s witnessed a dramatic growth in revenue from the tourism industry from 32 billion baht in 1985 to 110 billion baht in 1990. The visitors coming to Thailand rose from 2.5 million a year to 5 million in the second half of 1980s.[32] Thailand's position in the world shifted from an agricultural exporter to an industrial exporter, and an exotic tourist destination. Following Japan, investments from East Asia's newly industrial nations such as Singapore, Hong Kong and Taiwan also increased about tenfold, as the wage rate in these countries became high. Increasing revenue from exportation and influx investment from abroad ignited a boom in domestic consumption, which changed the retail pattern as small groceries in shop-houses at the corner of road intersections disappeared, and supermarkets, convenience stores, department stores and large-scale shopping malls mushroomed in urban areas. The government had committed itself to diversify investment to other provinces in the 1970s, but, in 1988, the National Economic and Social Development Board (NESDB) recommended that governmental policies focused increasingly on managing facilities in the Bangkok area. As a result, most industrial development projects took place in the outskirts of Bangkok and its neighboring provinces. Real estate development grew rapidly at a rate of about 20% a year from 1987 to 1990. The GDP of Bangkok Metro area became twice as high as the nation's GDP in 1980s. Bangkok also rose to share nearly half of the wealth of the entire nation while it took about 30% of the country in 1970s. The population of Bangkok Metropolitan area also grew at the rate of 7.3% a year and rose from 5 million to 9 million at the end of 1980s. Bangkok's green spaces, orchards and rice fields, in conjunction with the expansion of suburban residence, were gradually replaced by new factories, urban blocks of commercial shop-houses and housing projects.

With a growing influx of rural migrants searching for employment in the business and tourism sectors, traffic congestion in Bangkok became unbearable. The Expressway Authority was established in 1972 to cope with the growing traffic congestions, but it took ten years to complete the first phase of the expressway connecting the north and east suburb to downtown in 1981. The second phase finished in 1993, connecting the west side of Chao Phraya to Bangkok inner city. Moreover, BMA also cut a number of minor roads and added signage along alleys rather than connecting the main arteries together. The number of automobiles increased as a growing middle class emerged from the industrial boom. Thailand became the hub of the automobile industry in Asia. The first phase in the development of the industry was the assembling of cars and the production of parts in the 1970s–1980s for a domestic market. The industry then developed further by exporting entire vehicles in the 1990s.

The completion of the Expressway and BMA's effort to create shortcuts did not solve the traffic problem in the long term, on the contrary, it encouraged the use of automobiles and the expansion of housing development in the outskirts of the capital city. A greater number of cars causes more air pollution and heat gain. The growing suburban housing and real estate developments caused the deterioration of natural environments at the fringe of the metropolis. The green area was replaced by concrete surfaces thus increasing heat and reducing capacity of land to absorb flooding water. The expansion of public services such as electricity lines and clean water pipeline also increased energy consumption in the community.

The Bangkok political and administrative system was also solidified into a single unit to cope with the multiple expansions that had taken place since 1972. A governor was elected for the first time in 1975 but, with political instability in the latter part of 1970s, it was suspended. The governor election resumed in 1985 and triggered new developments in public facilities, in particular mass transit, urban parks and water treatment systems. A plan for a mass transit system had been initiated in the 1970s but sluggish to implement. The project was strongly backed when the elected governor, Major General Chamlong, came to power in 1985. The political conflict between the Bangkok governor and the Minister of Interior delayed the construction while the traffic situation in Bangkok deteriorated and threatened to damage Thailand's economic progress. Bangkokians prevailed by electing the same party's governors for three consecutive terms, making room for the construction of an elevated train. The Bangkok Transit System (BTS) finally opened in 1999 and completed the extension line to the western suburb in 2013. The first phase of BTS grabbed roughly 200,000 passengers from roads daily and the extension brought the ridership to passing 750,000 a day. The underground train (MRT) was opened in 2004 running through a loop of eastern rings connecting the northern and eastern suburb to downtown with a daily ridership of approximately 250,000. The total mass transit system has soothed Bangkok's traffic by taking a million passengers out of the road network daily.

In stark contrast to the traffic congestion problem, Bangkok in the 1980s also experienced an improvement of its urban parks. Prior to the 1960s, Bangkok had only three public parks. The largest and oldest urban park in Bangkok, Lumpini Park in downtown Bangkok, was 142.3 acres. It was originally designated as a site for the Thailand Exposition but converted to a park in 1925 after the plan for an international fair failed to realize. The other two parks are Saranrom Park (nine acres) and Thonburirom Park (25 acres), which were also converted from a palace in the historic district and royal land in the west. In 1980, 61-acre Chatuchak Park was created from the railroad yard northern suburb of Bangkok and 197-acre Rama IX Park, the largest public park in Bangkok, was established seven years later in the eastern suburb. From 1990 to present, another 27 public parks were established with smaller community gardens bringing the ratio of green area per capita of Bangkok to 6.25 square meter/person. However, the ratio is still very low at about 1–2 square meter/person in the historic city, downtown and inner suburban districts. With traffic congestion on the main streets, air pollution became potentially a serious public health problem in Bangkok.

Bangkok Looking West for the New Millennium

Moving forward into the new millennium, Bangkok emerged from the financial crisis in 1997 with upgraded facilities to cope with urban expansion but at the expense of the environment. For instance, the new international airport was opened in 2006 with its elevated rail line connecting to downtown. However, it was built on a reserved agricultural zone designated as a floodway in the east. New urban development such as hotels, apartments and housing estates were soon constructed blocking the runoff water to be drained into the Gulf of Thailand. The air traffic at the new airport has grown fast and is overcrowded, to the extent that the old international airport at Don Mueang was reopened in 2007.

The Department of Drainage and Sewerage (DDS) was founded in 1977 as a main authority to manage wastewater and prevent flooding. The first wastewater treatment plant started to operate at the inauguration of the DDS. The National Housing Authority also built 12 sewage treatment plants to serve public housing projects in suburban areas and all were also transferred to DDS from 1990 to 1997. From 1994 to 2006, seven large-scale treatment plants were constructed covering a total of 191.7 million square kilometers in Bangkok. Although wastewater treatment plants were constructed, waste kept being dumped into canals from households along the waterfront in western districts.

The large-scale permanent flood prevention system was initiated in 1984 to protect the inner city. DDS employed the polder system by embanking the area hit by rainfall and pumping flooding water out through main sewage pipe. There are 18 polder systems in 15 districts of Bangkok. From 1996 to 2010, the DDS also built a 77-kilometer-long flood dyke along the west bank of the Chao Phaya River and the canal loop of Bangkok Noi to protect 28 square kilometers of communities in Thonburi. From 1977 to 2010, the pumping stations grew from two to 150 stations with 685 machines. Floodgates and pumps were installed on the canals and river to control the level of water. The construction of seven main underground sewage tunnels was also completed and operated to channel floodwater to the river in 2009. The electricity cost for the pumping system and the controlling of floodgates was significant. The two-meter-high dyke damaged the traditional waterfront landscape and everyday life activities of the Bangkok community.

Open space, pond and swamp areas at the eastern and western outskirts were converted into retention basins to store 12.75 million cubic meters of storm water. Yet the east of Bangkok still needed additional storage of 6.25 million cubic meters to survive flooding. In addition, the storm water management and retention system has not yet been followed by building regulations for new urban developments. Every year, additional real estate developments have been built, replacing natural surfaces with impermeable pavement. The great flood of 2011 proved that all these high cost flood prevention systems failed.

The mass transit systems were under construction to move westward. The BTS sky train system completed a link to the southwestern suburbs and to Bangkok downtown, while the underground train MRT has expanded across the river to the west. The development of mass transit stations could in effect erase the last canal communities in the Bangkok region, because building regulation and land use planning offer incentives for locating large-scale commercial or residential projects close to transportation centers. A second ring road was constructed and completed as a freeway loop connecting the eastern and western outskirts of Bangkok. Housing developments followed the new road and soon caused traffic jams at the entrance of these projects. The commercial strip will soon increase the volume of storm water into the drainage system. The expressway also completed a loop to connect the southwest area of Bangkok to the industrial zone along the southeast coastal line, but it will soon be filled by trucks transporting goods from factories to distribution centers around Bangkok.

While most of the economic change from the 1950s to the new Millennium has drastically transformed the eastern regions of the Bangkok Metropolis, districts on the west side of the Chao Phraya River or Thonburi were marginally altered at a slower pace of development. New drainage and sewage protection laws largely safeguard the canal loop and network in the west side although it was not well maintained and some parts have become shallow and are blocking the water drainage pattern. Road networks in the western districts of Bangkok or Thonburi have not expanded greatly, thus still preserving old urban aquatic communities, market places and orchards along the canal loop. However, with the new extension of the mass transit system and freeway, soon new developments will follow and repeat similar growth patterns that have already made urban settlements vulnerable to natural disaster.

After the millennium, the urban space of Bangkok became overcrowded with the mega infrastructure of the BTS sky train system, elevated expressways overpassing communities, underground mass transit and high-rise towers in almost every district, reflecting the image of a globally recognized metropolis in Asia. Bangkok appears to be a fully modern city, but traffic is still congested in rush hours everyday, sidewalks are rutted, canals are filthy and floods still occur on major streets. The patterns of climate change bring heavy rainstorms that trigger flash floods, while high tides cause coastal land recession as seawater rises every year. Nature now represents an intimidation to human settlement and the security of Bangkok.

The Bangkok Municipality Administration, central government, Bangkok residents and academics should reconsider this conventional pattern of modern economic growth and explore alternative approaches to how Bangkok can reconcile itself with its natural environment with stricter environmental protection regulations and incentives for sustainable development. Communities along the old canal loop are now reinventing themselves and becoming sightseeing spots. The early urban yet aquatic settlements gained more recognition from Bangkok residents as the waterfront markets along the canals of Bangkok and neighboring provinces became popular. Communities should be supported by the policies that encourage the operation of water transportation and remove obstacles in the canal system. Canal transportation could reduce the traffic congestion and provide an alternative mode for traveling to the outskirts of Bangkok. More commercial activities would soon follow the node of canals and land-based transportation and could connect to the network of waterfront markets. Therefore old canals could gain more public attention and better maintenance.

Bangkok also needs large areas of water retention basins in order to prevent flooding. These areas can also be utilized as public open spaces and recreational facilities and should be integrated into a public park system. Retention areas with storm water management connecting to canal systems and the river can become a network of open space bringing urban users to the nature of a water-based city again. For new developments, the urban design policies should provide incentives to design the ground floor as open or green space thus allowing water to flow. The open space on the ground floor should be substituted with extra space and the height limit be made more flexible. The building regulations should set the requirements for mega-projects so that they retain water from rainfalls, before being released into the public drainage system. The owner could also retain water in longer periods to use in times of drought, which also usually occurs. The state agencies, businesses and residents of the Bangkok metropolitan region from now on have to find the solution in their lives to reconnect their living pattern with an awareness of the major natural elements of the lower river basin, which are water and its network.

Notes

1 Piyanart Bunnak, Wat nai Krung Thep: kanplianplaeng nai ro_p 200 pi Pho. So. 2325–2525 (Buddhist temples in Bangkok 1782 A. D.–1982 A. D.) (Bangkok: Chulalongkorn University, 1982), 124.
2 Pasuk and Baker, *Thailand: Economy and Politics* (Singapore: Oxford University Press, 1997), 243–255.
3 After political turmoil following the coup in 1932, the military leader completely took control of the government in 1938 and last until 1944. See more in Marc Askew, *Bangkok Place, Practice and Representation* (New York: Routledge, 2002), 47, and see more for the population growth in Larry Sternstein, *Portrait of Bangkok: Essays in Honour of the Bicentennial of the Capital of Thailand*, (Bangkok: Bangkok Metropolitan Administration, 1982), 94.
4 After a brief encounter, Thailand agreed to ally with Japan, avoiding occupation by the Japanese army, however, there were some damages in key strategic areas of Bangkok. See more from Thepchoo Thubthong.
5 The ruling elites first referred to the royal rulers who managed to escape direct colonial rule by complying with colonial demand for free trade and adopting colonial apparatus in transforming the kingdom into a modern nation. However, they did not survive the world's depression in the 1930s, which led to the end of the absolute monarchy. Ruling elite in this case also refers to those in the network of royalists and military leaders that controlled Thailand.
6 Pasuk and Baker, *Thailand: Economy and Politics* (Singapore: Oxford University Press, 1997), 125.
7 Ibid., 126–127.
8 Ibid., 279.
9 Maurizio Peleggi, *The Politics of Ruins and the Business of Nostalgia* (Bangkok: White Lotus, 2002), 61–62.
10 Ibid., 62.
11 https://th.wikipedia.org/wiki/ท่าอากาศยานดอนเมือง
12 Peleggi, *The Politics of Ruins*, 62.
13 Wolf Donner, *The Five Faces of Thailand: An Economic Geography* (St. Lucia: University of Queensland, 1978), 836.
14 Marc Askew, *Bangkok Place, Practice and Representation* (New York: Routledge, 2002), 53 and Pasuk and Baker, *Thailand: Economy and Politics* (Singapore: Oxford University Press, 1997), 128.

15 Marc Askew, 53.
16 Pasuk and Baker, 128.
17 Steve van Beek, 137.
18 Ibid., 141.
19 www.port.co.th/sitenew/en/topic1_know01.php
20 Thai University Research Associates, *Urbanization in the Bangkok Central Region* (Bangkok: Social Science Association of Thailand, 1976), 268.
21 Department of Town and Country Planning, *Report on the First Revision of the Plan for the Metropolitan Area Bangkok, 1971* (Bangkok: Planning Division, 1972).
22 Nipan Vichiennoi, "Urban Housing in Developing Countries: A Case Study of the Socio-Economic Status of Housing Purchasers of New Housing Estates in Bangkok, Thailand 1969–1979," PhD Dissertation, Oxford Polytechnic, 152–154.
23 Marc Askew, *Place Practice and Representation* (London: Routledge, 2002), 78.
24 Larry Sternstein, *Portrait of Bangkok: Essays in Honour of the Bicentennial of the Capital of Thailand* (Bangkok: Bangkok Metropolitan Administration, 1982), 91, citing map by Chira Chitrakorn of City Planning Department of the Bangkok Metropolitan Administration based on land use map by the Department of Town and Country Planning, Ministry of Interior.
25 Anuchart Poungsomlee and Helen Ross, *Impacts of Modernization and Urbanization in Bangkok: An Integrative Ecological and Biosocial Study*, (Nakhon Prathom: Institute for Population and Social Research, Mahidol University, 1992), 19.
26 Steve Van Beek, *The Chao Phya River in Transition* (Kuala Lumpur: Oxford University Press, 1995),162.
27 Department of Drainage and Sewerage, *Bangkok . . . a City by the Water* (Bangkok: Bangkok Metropolitan Administration, 2010), 86.
28 Steve Van Beek, *The Chao Phya River in Transition* (Kuala Lumpur: Oxford University Press, 1995), 163.
29 Ibid.,164.
30 Anat Arbhabhirama, Dhira Phantumvanit, John Elkingon, and Phaitoon Ingkasuwan, *Thailand Natural Resource Profile* (Singapore: Oxford University Press,1988), 137–138.
31 Pasuk, 148.
32 Ibid., 153.

References

Arbhabhirama, Anat. Dhira Phantumvanit, John Elkingon, and Phaitoon Ingkasuwan. *Thailand Natural Resource Profile*. Singapore: Oxford University Press, 1988.

Poungsomlee, Anuchart, and Helen Ross. *Impacts of Modernization and Urbanization in Bangkok: An Integrative Ecological and Biosocial Study*. Nakhon Prathom: Institute for Population and Social Research, Mahidol University, 1992.

Askew, Marc. *Bangkok, Place, Practice and Representation*. London and New York: Routledge, 2002.

Askew, Marc. "Transformation and Continuities: Sacralization, Place, and Memory in Contemporary." in Ronald A. Lukens-Bull (ed.), *Sacred Places and Modern Landscapes: Sacred Geography and Social-Religious Transformations in South and Southeast Asia*. Tempe: The Program for Southeast Asian Studies, ASU, 2003.

Department of Town and Country Planning. *Report on the First Revision of the Plan for the Metropolitan Area Bangkok, 1971*. Bangkok: Planning Division, 1972.

Department of Drainage and Sewerage. *Bangkok . . . a City by the Water*, 86. Bangkok: Bangkok Metropolitan Administration, 2010.

Donner, Wolf. *The Five Faces of Thailand: An Economic Geography*. St. Lucia: University of Queensland, 1978.

Vichiennoi, Nipan. "Urban Housing in Developing Countries: A Case Study of the Socio-Economic Status of Housing Purchasers of New Housing Estates in Bangkok, Thailand 1969–1979." Ph.D. Dissertation, Oxford Polytechnic,1982.

Phongpaichit, Pasuk, and Chris Baker. *Thailand Economy and Politics*. New York: Oxford University Press, 1995.

Peleggi, Maurizio. *The Politics of Ruins and the Business of Nostalgia*. Bangkok: White Lotus Press, 2002.

Bunnak, Piyanart. *Wat nai Krung Thep: kanplianplaeng nai rop 200 pi Pho So 2325–2525* (Buddhist temples in Bangkok 1782 A. D.–1982 A. D.). Bangkok: Chulalongkorn University, 1982.

Sternstein, Larry. *Portrait of Bangkok*. Bangkok: Bangkok Metropolitan Administration, 1982.

Kan-Ari, Thammakiat. *Phranakhon Khuan Chom*. Bangkok: Manager Group, 1995.

Van Beek, Steve. *The Chao Phya: River in Transition*. Kuala Lumpur: Oxford University Press, 1995.

Retrieved from https://th.wikipedia.org/wiki/ท่าอากาศยานดอนเมือง.

Retrieved on June 18, 2013 from www.port.co.th/sitenew/en/topic1_know01.php.

3 Growing Up in a Steel Mill

A Conversation with Historian and Nature-Lover Elmer J. Hall

Michelle L. Stefano

It could be said that when one thinks of the countless mills, factories, and manufacturing plants across the US, and indeed the world, *nature*—in terms of how it is most popularly conceptualized—does not usually come to mind. One may think of large, imposing factory buildings made of brick and/or metal lined up in rows, or spotting wide expanses of treeless landscapes. The soundscapes can be considered *unnatural*, too, with the belching of smoke stacks, the loud grinding of machinery, and in earlier days, the frequent sounds of trains transporting raw materials and goods, among other noises. On the path of industrial boom over the 19th and 20th centuries, these centers of manufacturing grew to encompass not only sites of work for millions of people, but also their homes, forming distinctive ways of life. Thousands of company towns were built to house workers and their families, and to provide the amenities needed for self-sufficiency, all under company control.[1] Over the first half of the 20th century, these towns expanded beyond company borders and an increase in industrial jobs helped to fuel the rapid development of a significant number of American cities. In this light, when thinking of the places and spaces where the lives of an immense workforce were shaped, it is possible that nature was there, and that it played a role.

This was true for Elmer J. Hall, who was born into a life heavily influenced by industry and, more importantly, the relationship—both interconnections and contrasts—between it and nature. Born in 1942, Elmer grew up in the company town in the middle of the Sparrows Point Steel Mill, located on the deep-water port just outside of Baltimore City, Maryland. His father was a mill worker, which was a requirement for living in the town, and his mother was a homemaker. From the 1940s through the 1960s, Elmer went to school, played sports, fished and swam all within the confines of one of the largest steel mills in the world.

Sparrows Point, as it is known, employed over 30,000 employees at its peak in the mid-20th century.[2] From 1887 to its closure in 2012, the lives of hundreds of thousands of steelworkers and associated personnel were affected by mill operations, expansion, and for many to this day, its eventual shutdown. Particularly impacted by the mill on a daily basis was a subset of this population—roughly 9,000 people at its height before the Great Depression[3]—who lived in the company town between the 1890s and 1972. Named after the mill, the town was situated in the center of the Sparrows Point peninsula and surrounded on three sides by the steelworks and with many points of access to bordering waterways.

As Deborah Rudacille describes: "Work, family, and community were tightly braided together on Sparrows Point, where an employee needed a letter from his foreman to rent a house and where three generations of some families lived surrounded by coke ovens, open hearth furnaces, rolling mills, and enormous piles of limestone and coal."[4] For most of its existence, the company town of Sparrows Point functioned as a self-sufficient village, with its own police force and fire department, schools, places of worship, stores and recreational activities, among many other amenities. Significantly, the presence of water—the surrounding harbor, river and creeks—provided ample opportunities to engage with nature, in its traditional sense, as did the various green spaces, such as sports fields and marshlands found in and around the town.

In 1972, Bethlehem Steel, the longstanding company that owned the mill until the early 2000s, forced residents to leave in order to raze the town and make room for the "L-furnace," the mill's biggest blast furnace[5] and, at the time when it became operational, the largest in the Western Hemisphere.[6] Homes were destroyed, and almost all traces of the once-thriving town were disappeared forever; although, traces of the company town streets can still be viewed on Google Maps (see Figure 3.1). Elmer is one of many people who not only lost a neighborhood to which he could no longer return, but also the places and spaces in which his childhood experiences and memories were formed and remain grounded.

It is of no surprise that Elmer, who became an educator and social worker, turned to documenting the oral testimonies of those who lived and worked at the "the Point" over the past decade, years before the mill's closure. In his first book, *Diary of a Mill Town: Recollections of the Bungalows and Sparrows Point, Maryland*, which was published in 2004, he brings together the memories and stories of numerous residents, structured in relation to particular places, such as the town's churches, stores, public spaces, as well as associated annual events. For instance, Fran Baker Keil recalls her time fishing in "the Bungalows" area of the town, where Elmer grew up, and states:

> There were several piers on Haddaway Road. All of them built by residents of the Bungalows. Heffler's had a nice pier. Next to it was Paw Ashton's, Fritz Baker's, and then Mac McGee. We kept our small boat there. It was a great place for summertime fun. Lots of crabbing and fishing. Seafood was plentiful.[7]

Elmer also has strong memories of crabbing and fishing off of these same piers, as recounted later. Nonetheless, these activities were most likely hazardous to residents' health. He has since realized how polluted the waterways surrounding the mill must have been and, indeed, are today; "it's a wonder that we don't glow," he jokes. His other four books, all dedicated to Sparrows Point history, are as follows: *Shipbuilding at the Sparrows Point Yard: A Century of Pride and Tradition* (2007); *The Patapsco and Back Rivers Railroad: Chronicles of the Push, Bump and Ram* (2010); *A Mill on the Point: 125 Years of Steel Making at Sparrows Point, Maryland* (2013); and the forthcoming, *The No. 26 Street Car Line: Chronicles of the Red Rocket.*

In early 2017, I was lucky to sit down with Elmer to gain insight into his steel-town upbringing and the sorts of experiences with nature he had had. The following interview has been edited with his permission.

Michelle Stefano: So, Elmer, please introduce yourself.

Elmer Hall: I was born and raised in the company town of Sparrows Point, Maryland. It was located in the very heart of a gigantic steel mill, the Sparrows Point Steel Mill. The town itself dates back to 1887 when Pennsylvania Steel Company bought the peninsula of Sparrows Point and decided to build a rail mill there. A man

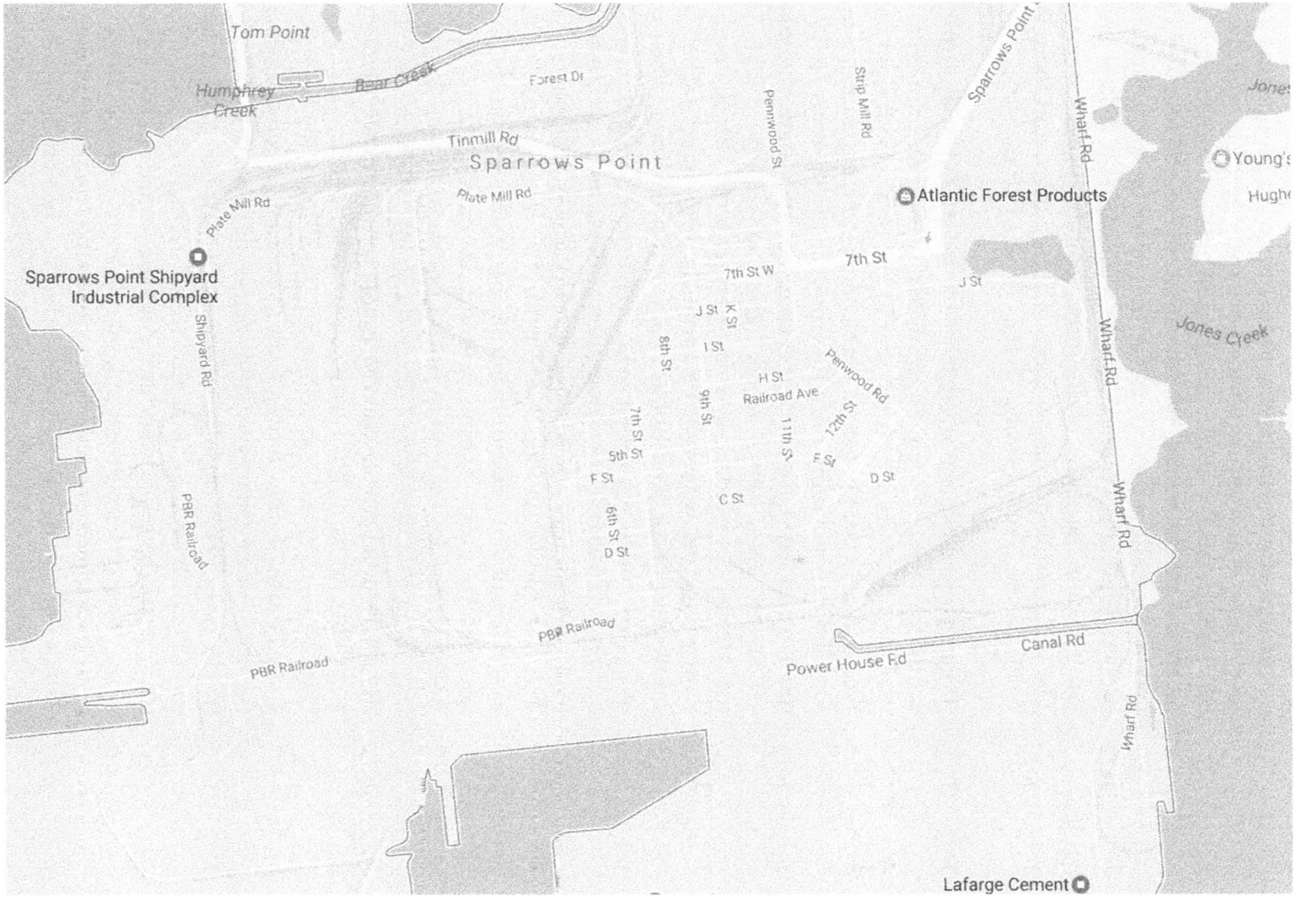

Figure 3.1 Sparrows Point, Google Map, 2017

named Frederick Wood and his brother, Rufus, came from Lowell, Massachusetts and built the town. Frederick built the steel mill and his brother laid out the town, and they laid it out in a grid fashion after their town in Lowell. One of the nice aspects of the Sparrows Point town with respect to nature was that every street was lined with sycamore trees, so there was always this lush greenness in the midst of the stark contrast of black metal and smoke stacks. Why sycamores? I don't know, because they're basically a river tree, but they were on every block, on every street on the Point. The town itself was really like a small, green oasis in the center of this large, looming gigantic steel mill. The mill surrounded us on three different sides, and there was this little speck of greenness in the center of all that.

MS: Where exactly did you live?

EH: From east to west, the streets were lettered, and from north to south, the streets were numbered. My streets were not laid out that way; the section of town that I lived in was built right after World War I and, I guess, as Bethlehem [Steel] expanded, which they did exponentially once they bought the property in 1916, there was a need for more housing. So, they built this place called "the Bungalows," where I grew up [see Figure 3.2]. And the street that I lived on was named after a tree: Beechwood.

MS: I am wondering what you meant in the beginning of The Bungalows section of your book, *Diary of a Mill Town*, when you note: "Although we are all from Sparrows Point, and consider ourselves 'Pointers,' our geographical detachment did foster, in many of us, a separate sense of identity as a community"?

EH: The Bungalows was on the outskirts of the Sparrows Point town. It had a reputation from people who lived on the Point not that far away—within walking distance—as a place that was less than desirable to live. And I'm not sure why. I guess, it was about that they were "up on the Point" and you were literally "across the tracks" [in the Bungalows], and we were poor and didn't know it, but I think we were better off.

Here's an example: I was selling my book about the Bungalows at the Fourth of July Heritage Fair in Dundalk [neighborhood to the north of the Sparrows Point peninsula], and a woman who lived on the Point came by and said, "Oh, a book about Sparrows Point. It says the Bungalows on there?" And I said, "Well, that's what the book is about." And she leaned over to me and said, "We weren't allowed to go down there." I said, "I'm sorry to hear that, but you missed a wonderful opportunity to see a wonderful community." And I didn't sell her on it, but it had this reputation undeservedly so. But we had our own community organization, separate from Sparrows Point. They had a playground, a ball field . . . we had our own Halloween parade. At Christmas time, we had our own Santa Claus who would go around on a jeep that looked like a sleigh with two elves in the

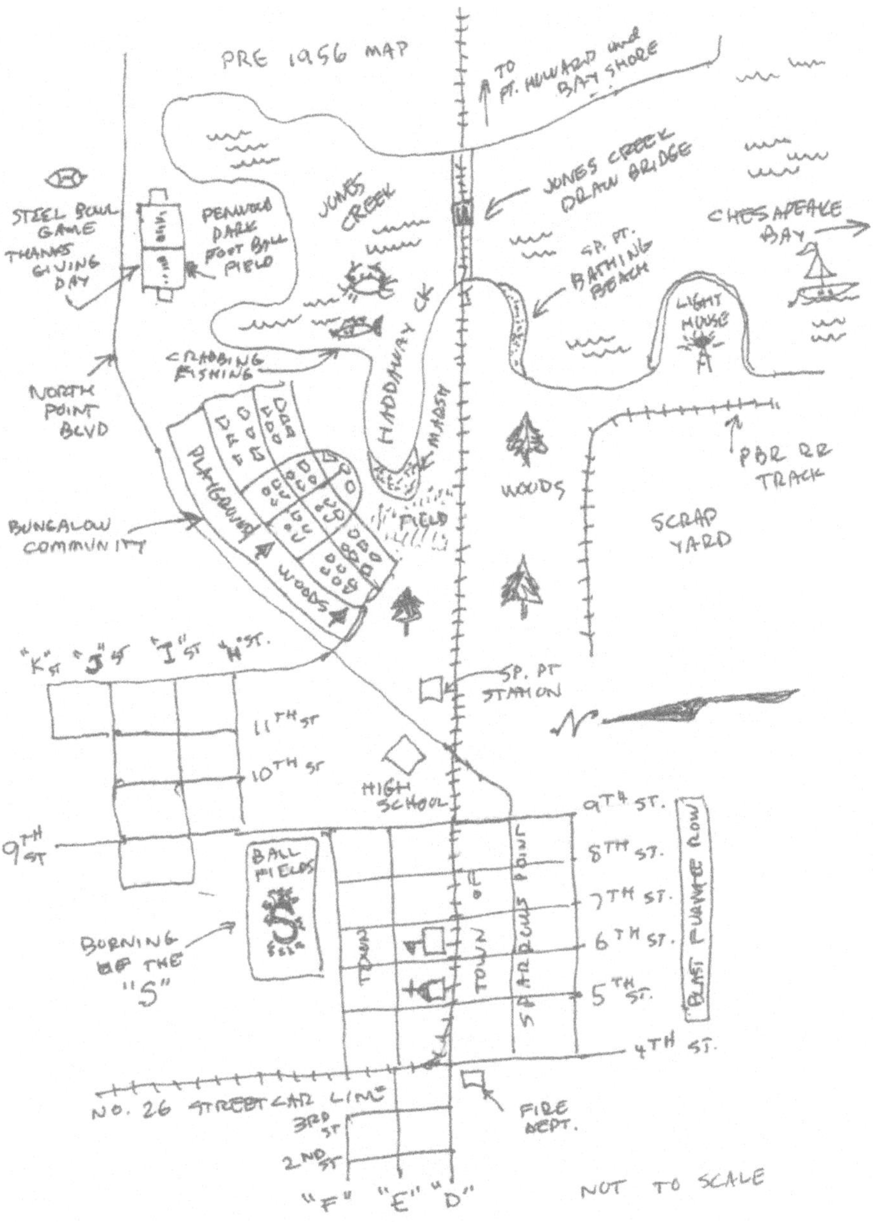

Figure 3.2 Map of the Sparrows Point Steel Mill Company Town Drawn by Elmer J. Hall, 2017

front—and they would ride through the three streets in the Bungalows area, handing out candy to every family. And we were the only community with waterfront property, so we had the piers and the boats, crabbing and fishing. None of that existed up on the Point. We had our own identity.

MS: What was the Point like?

EH: Up on the Point, in the main part of the town, the large houses on B Street belonged to company supervisors, and there were three mansions to a block. Compared to what I lived in, some of those houses had three or four stories, fifteen rooms, and five bathrooms, walnut staircases, fireplaces . . . they were opulent. The house that I lived in had 500 square-feet, and that's not a whole lot. From D Street and that point on [going down the alphabet], you could live wherever you want, it was not necessarily categorized by your job level as most people usually think. The main street was D Street and that's where the Red Rocket [streetcar, or trolley, to and from Baltimore City] came in on, which was our main source of transportation in those days as we did not own a car. As you came onto D Street, most of the churches—there were seven of them in the town—were located there. Five were located in the white community and two in the black community. When I was growing up, the town was segregated—the blacks lived on the north side, and the whites lived on the main side of town . . . separate schools, separate everything. There were a couple of drug stores [pharmacies] on the main street, a couple of restaurants, there were two banks, the elementary school, the telephone office, Caplan's Department Store, which was where I bought all my school clothes, Mr. Bateman's supermarket, a shoe repair shop, a dentist, and two doctors' offices. It was pretty much a self-sufficient town.

Up until 1944—and I was born in 1942—there was a company store, which was like a Walmart way before its time: everything that you could ever possibly want—from a thumbtack to a tuxedo—they had it. But it was on a scrip service. People weren't making a lot of money, but you would get these scrip books, and you'd go to the store and they'd subtract what you bought from your pay. You were really living from paycheck to paycheck on that scrip system. I talked to Mr. Caplan [years later] and he was happy when they no longer continued the company store system because then his business boomed . . .

MS: And what about the schools?

EH: Sparrows Point was one of the first places to have a kindergarten south of the Mason-Dixon Line[8] because the brother of the man who ran the town believed that the steel workers would be better if they were educated. It opened in, I think, 1892 and I went to that actual kindergarten back in the mid-1940s. But it was company built and operated, and they built the elementary school in 1903 once the town started to expand. And they took that little [original] elementary school that later became a drug store on D Street, and on the 700 block of D Street, they built a two-story, 12-room school for the first to sixth grades. From where I lived in the outskirts of town, I guess it was about a little less than a mile that I would have to walk.

So, we had a kindergarten, an elementary school and a high school, and we had an African American elementary school. The African American kids didn't have a high school until the late 1940s or early 1950s or so, but the black children couldn't go to our high school until the Civil Rights Act of 1964. They would have to catch the streetcar, or a bus, into town [Baltimore City], if they wanted to get a high school education. Now, the African American part of town was small; they lived on I and J Streets, which had about six to eight blocks each on them, and two blocks on K Street. It was a very small community. Louis Diggs[9] notes that the black population living in Sparrows Point had the highest percentage of professionals to come out of that community, with degrees in medicine, nursing, higher education or law. And it's possible that this was due to the fact that the lowest jobs on the totem pole at the mill were those filled by African Americans.

I mean, discrimination was rampant, but growing up, you just accepted it—you didn't know any better. We had a bathing beach, which was just a beach with no lifeguard or anything, and every once in a while, you would meet—coming along the streetcar tracks—the African American kids. Now, they could not swim at the beach, which was somewhat a public beach with a little stand that you could buy sodas and stuff at, but they had to swim at another beach. Theirs was not even a beach; it was just on the left side of the Jones Creek Bridge. But we never questioned that, you know? I mean, it just never came up. And I wasn't raised in a home where my parents expressed any kind of prejudice. It just was never explained. It's kind of weird.

MS: What were some of the social activities for residents, especially for the youth?

EH: In terms of activities, the company was very good about doing things for kids. They had large areas for ball fields—mainly for baseball and [American] football. When I was growing up, there were no football teams in the high schools, so we played football for the YMCA [Young Men's Christian Association]. And the biggest event that we had was on Thanksgiving Day, on Penwood Field, and it was called the "Steel Bowl Game." And Sparrows Point would play the Dundalk team. And the rivalry between the two still exist today when someone says they grew up in Sparrows Point. We know the exact border between the two [areas] and we would say, "No, you're not! You're not a Pointer!" So, there was always a strong rivalry. There were three teams based on age groups and we played three games back to back. And the night before the game, there was a big Thanksgiving Day parade led by the fire department—and Sparrows Point had its own fire and police departments run by the company. So, the fire trucks would start in the Bungalows, where I lived, and they had flatbed trucks that the football team would get on, and we would ride through the town, even through the north side of town, zig-zagging through the entire town. And we would end up in the field behind F Street, where everyone was gathered and we would burn this gigantic "S," with the fire department's supervision. A large "S" for Sparrows Point, and that would be the culminating event of the celebration on the eve of Thanksgiving, and then we'd compete on Thanksgiving Day . . . and then there was a dance after the games . . .

MS: What was it like on a day-to-day basis growing up there?

EH: Both of my parents grew up in the foothills of the Blue Ridge Mountains in Virginia; they came up during the Great Depression to find work. My father worked in the shipyard; he was a chipper there, and my mother was a housewife. So, they brought with them some aspect, I guess, of the "backwoods" . . .

MS: How's that?

EH: Well, I just think that they grew up on farms and they were used to hunting and agriculture and all that kind of stuff. Like, my father would come home every day from the shipyard, and he had a routine: he would come home, he would take a hot bath, he would eat supper, and after supper, we would take a walk. And we would walk through the woods, cross the streetcar tracks, through another wooded area, follow the railroad tracks to this place that we called "the lighthouse," which is still there—the lighthouse is still there. And the lighthouse was on this little peninsula, surrounded by beach on three sides and scrap yards on the other side. And for a young boy, it was just a magical place to go. We would beachcomb and find things like . . . I guess a real find would be a dead duck, because there was always this opportunity that they had a [metal] band on its leg, and that was just like a real find . . . that was like finding a treasure trove . . . if you found a banded duck, you'd collect the band. There were crab buoys and all kinds of stuff to explore on the beach, but in the center of it was this wooded area with rabbits, squirrels, bears . . . not bears [laughs] . . . birds, squirrels, rabbits, quails . . . all those small game animals. It was one of my favorite places to go . . . one of my favorite childhood memories was to visit the old lighthouse site.

MS: For those who may not know, what is a crab buoy?

EH: Well, a crab buoy is . . . if you're a commercial crabber, you had these crab pots. They were, I guess, two and a half feet by two and a half feet; it was like a cube. And it had and opening on it and inside it, you placed bait—some kind of smelly fish. And they would have, I guess, 20, 30, 40 of them sometimes, and they took them out and placed them in areas of 8, 9, 10 feet of water. And to mark where you dump the crab pot over in the water, they had a cord to it with a buoy. Each marker was painted a distinctive color to identify the owner; it had a number on it and a color on it. Sometimes, boats would come by and hit those cords and the crab buoys would float away . . . and the beachcombers would find them.

MS: Now, taking into account that these waterways surrounded one of the largest steel mills in the world, I would think that they were really polluted . . .

EH: Well, I think the water was polluted, and it's a wonder that most of the people who are my age—and I was 75 yesterday—it's a wonder that we don't glow. But I think there was a lot of indiscriminate dumping—in the early days—of toxic acids and things like that, you know, all kinds . . . because it was just a readily available source to just . . . it was easy to dump; there were no regulations in those days. I know for a fact that the sewage treatment plant was located in the Bungalows. There's still a pipe located out in Jones Creek where I'm sure the waste was not as refined as it should have been; raw sewage was also dumped in addition to the toxic waste from the steel mill. They cleaned that up a lot in the 1960s, especially the sewage, but also the highly toxic stuff—they really cleaned that up. But with the rules and regulations about emissions and, you know, clean water . . .

MS: But while growing up, of course, you were not aware and you swam in the water?

EH: But growing up we would swim in it; we would eat the fish . . . we would crab. Right behind the house we had . . . actually, we were surrounded . . . the Sparrows Point peninsula was surrounded by water on three sides. On the east, it was Jones Creek; on the south, it was the Chesapeake Bay; on the west was the Patapsco River; and on the north was Bear Creek, so we were surrounded by water.

You know, there was a lot of beach area to explore; boats were everywhere. In the Bungalows area, the men were allowed to build their own piers to moor their boats. So, there was always access to the piers and we would crab and fish off of them. We would go up to Mr. Bateman's—it was called a supermarket, even then in the 1940s and 1950s—and buy chicken heads and feet and we would take them home and tie about a ten-foot string to each of them. Then, we would slip a washer or a [metal] nut on it so it would sink, and we would take those out to the piers and we would crab. You take the line and tie it to a pole, and we would have about ten of these lines, and you would drop them into the water and you would sit and wait. And a crab would come along and want to take the piece of chicken some place to eat it. And you watched your line—and we called it "stretchers"—and as your line went straight out at an angle in the water, you knew a crab was pulling it; it was stretching. You'd get on your hands and knees, you would have a net to your right, and you slowly inched the line in as close as you can, because if you pulled too fast, he would drop off, so you pull and pull, inch by inch, and then you scoop up your crab and put it in a basket. And if you get a dozen or two dozen, or a bushel, you'd take them home and steam them.

MS: And how did you learn that?

EH: Everybody did it; it was part of the culture. You know, crabs were there and for the taking. I'm not sure where I learned it, probably from other kids, but we all did it. And when we weren't crabbing, we were fishing, and when we weren't fishing, we were swimming. We also had a cat-tail marsh that adjoined the Haddaway Creek, and that was great to explore, too. I was just a nature freak; I just loved the outdoors and I could not get enough of it. And I think it came from those early days with my father, taking those walks when he would get home from work. But I just loved to put a pair of boots on, mucking through the cat-tails. As you would go through, you would see red-wing black birds and they would hover up, and you knew you were near a nest. And, of course, if it had eggs in it, it was something I'd go back to check and watch for them hatch, watch them develop and take off. Every once in a while, you would stumble upon a large snapping turtle the size of a bushel basket. For an eight-year-old kid, dragging something like that by the tail was a real prize. But not so much for the snapping turtle, because he ended up in a pot [laughs]—we made turtle soup out of it.

And right behind my house on Beechwood Road, we had a field that I was always exploring. There were pieces of tin, cardboard, logs, and every single time that I went there, I could lift up these things and there was always, always a snake under it. They were small, brown snakes—called DeKay snakes. Well, sometimes I would take one and stick it in my pocket and forget about it. And when my mother would wash my clothes, you could hear the screams two blocks away as she discovered them! Boy, did I get in trouble . . .

MS: I'm curious about how the steel mill influenced your day-to-day life, because you're also in the middle of it . . .

EH: I mean, the steel mill was huge . . . the blast furnaces along Blast Furnace Row, which was in sight of my house, and there were like ten of them and 300 feet high. There were constant sounds of the furnaces, the trains clanging, pipes hissing, pile drivers going 24 hours a day. We just got used to the noises, and we got used to the sight of the mills that we were surrounded by all the time. That's probably why we looked for the greenery, or nature part of it, because of the stark contrast of the mills. I mean, we grew up with the pollution. You know, in those days, there were no emissions regulations, so if my mother was going to hang out clothes, she'd wet her finger and raise it into the wind to see which way it was blowing, because if it was blowing the way the dust was blowing, she wasn't going to hang the clothes on the clothesline.

I would visit my grandfather on his farm in Virginia and I'd have trouble sleeping because it was so quiet! I was so used to the constant noise of the mill. They would just dump molten slag[10] into the water—probably half of Sparrows Point today is landfill. And when the molten slag hit the water, there was a huge explosion. Houses, windows shook . . . the china in the cabinets shook . . . the sky lit up . . . and we grew up with that on a daily basis. It sounds like a warzone [laughs]! And I think that's probably why nature was a big part of my life, because of that contrast.

The houses where we grew up were owned by the company, and they maintained them. They changed the shingles on your roof every ten years, they painted them every five years, they'd fix plumbing issues . . . they owned everything and you owned nothing. You paid rent to live there. In order to live there, you had to work for the company. You owed your soul to the company store, kind of thing. However, they let you maintain your yard, and my mother had flowers everywhere to counter the dismal dark metal surroundings and it was somewhere to escape it. To this day, I still overkill my gardens with flowers and I think it stems from growing up there and my mom.

MS: What changed as you grew up, after the 1940s?

EH: I graduated high school there in 1960, and really, the town hadn't changed a whole lot. It was in its heyday at that point—as a matter of fact, the mill was growing. I didn't experience a lot of change until the 1970s, and my parents were still living there then, when the company decided that they were going to build the L-furnace, and they didn't have space for it. Because we were surrounded by water, and they had expanded to every place they could, the green oasis of the town became prime real estate and they evicted the entire population of 5,000 people, my parents included. And they did it without any safety net or counseling of any kind; they just said, "Get out." And this was in 1972. Everyone had a 30-day lease, and even Mr. Caplan knew that since they never gave a lease for longer than 30 days, some day they were going to come and tell you to get out. And it was a real hardship for my parents because they were paying 23 [USD] per month and the cheapest they could find in neighboring Dundalk was 125 [USD] per month, not including utilities. Talk about change—that was a big change.

MS: Did people fight it?

EH: It was almost like Tiananmen Square; they didn't stand in front of the bulldozers, but some people resisted until the last minute. When I was researching my first book, I found some photographs of the Warhurst family and they stayed until—you know, it was desolation, there was nothing left—the last house standing. You couldn't really resist it because your lease was up. It took about two years for everybody to finally get out and then they quickly came in and leveled the place. A couple of the houses were sold for a dollar, but you had to move them yourself. There are still a couple of those [relocated] houses around. But that was the end of Sparrows Point as I knew it. And they built the L-furnace and, I guess, in about ten years they started having problems with the economy, and the steel mill eventually deteriorated for good.

MS: How do you feel about industrial decline, a topic that is still so relevant today, especially in terms of its wide-ranging impacts?

EH: Well, at one time, Sparrows Point was the largest employer in the state [of Maryland]. In the mid-20th century, there were over 30,000 people employed there, and that included the shipyard, too. And it was like, "this will never disappear." You can talk to people and they would think that this was never going to happen . . . and it didn't turn out that way. And a lot of people are still devastated. Not only did I lose every aspect of the place where I grew up . . . you know, my holy communion, my first kiss in the fourth grade, all of our churches and schools, all of my childhood memories. I mean, all of that has been completely erased.

And the people who worked there are still having problems, because a lot of them were not quite at retirement age when it closed, but even the ones who did . . . my wife had 42 years there. She worked in the accounting department, in the main office. She had a good job . . . hardly ever missed a day except for her pregnancy . . . I mean, snow days, whatever, I got her there. She was totally dedicated to that place. Well, when she did retire, it was before [Bethlehem Steel] went bankrupt, but once that happened, a company named International Steel Group (ISG) came in under Wilbur Ross.[11] And Ross was the first owner [in a succession of owners until the mill finally closed in 2012], and he said that he'd buy the assets of the mill, but not the liabilities. And the liabilities were workers' pensions and health care. So, that disappeared for a lot of people—completely gone. Well, the government picked it up—the Pensions Guarantee Fund—and they didn't pick up health care, just the pensions, but they said [to my wife]: "You worked here 42 years, Mrs. Hall, but we're not going to pay you the pension amount you earned." So, her pension was pretty much cut in half. And that was hard to take. I had health care [from a separate job], so I brought her onto that with me, but should I predecease her, she'd be back to no health care after a year. There were some hardships there.

And the entire peninsula has been completely leveled. Every aspect of the steel mill is gone—it looks like Hiroshima. When they came in a couple of years ago to dismantle the place, they took it right down to the ground. When I was doing research on one of my books, I talked to people who said, "You really don't want to know what they did with all that pickler acid."[12] Environmental regulations or not, you don't want to know. But it was dumped indiscriminately, and I cannot imagine how they could build anything on the land. I hope they don't do anything residential, because I see Love Canal.[13] There was just tremendous abuses made with the toxic dumping, both in the water and on the land. It's a wonder that I don't glow from being exposed to all that. But they're going ahead; they've already started to build on it now. Fed-Ex[14] has moved in and has a big facility there already; Under Armour[15] is building a big distribution center there. I saw something in the newspaper, the *Dundalk Eagle*, that they're going to put a shopping center there. I just think if you take a look at how polluted it was—both the water and peninsula—somewhere along the line, it just seems that if there's enough politicians involved and enough money gets thrown around, somebody's going to say: "I don't smell anything." And that's just sad.

MS: When you look back, what feelings do you have about your childhood and growing up in the middle of one of the largest and most well-known steel mills in the world?

EH: There was something special that happened while we were growing up there. The friends that I have today are the same people I went to elementary school with there. They are my very best friends and there are ten of us. And we get together three times a year. Most people cannot recall who they went to elementary school with . . . or even high school classmates they don't even associate with today, but we grew up in the same houses, in the same place . . . and nothing changed while we were there, and some sort of bond grew. We cannot explain it. But that's what's special to me. Whatever happened in that little company town with them, there's still this strong attachment among us. Maybe we are all we have left of this place, which is true. So, we sit around—usually when we meet, around a campfire—and it's like we have our own language. We have these oral histories, stories that we sit around and share. Like, remembering Stuart McGee crawling around the floor in fourth grade class, shooting spitballs at Joanne Taylor . . . and we're keeping the place alive through our reminiscences. And when we get together, our wives go off thinking, "Oh, boy, here we go again—the same old stories!" We've been doing this for 20 years . . .

Notes

1 Hardy Green, *The Company Town: The Industrial Edens and Satanic Mills That Shaped the American Economy* (New York: Basic Books, 2010), 3.
2 Mark Reutter, *Sparrows Point: Making Steel—the Rise and Ruin of American Industrial Might* (New York: Summit Books, 1988); Deborah Rudacille, *Roots of Steel: Boom and Bust in an American Mill Town* (New York: Pantheon, 2010).
3 Rudacille, *Roots of Steel*, 130.
4 Rudacille, *Roots of Steel*, 6.
5 Blast furnaces are used to smelt iron ore to make "pig iron," a key intermediary step in the production of steel.
6 Reutter, *Sparrows Point*, 430.
7 In Elmer J. Hall, *Diary of a Mill Town: Recollections of the Bungalows and Sparrows Point, Maryland* (self-published, revised edition, 2004), 38.
8 The border between the states of Maryland, Pennsylvania and Delaware, set in the mid-18th century, which is commonly understood as culturally demarcating the border between the US North and South.
9 Louis Diggs is a historian who has studied and published extensively on African American communities in and around Baltimore, Maryland.
10 The waste produced from smelting iron ore.
11 As of March, 2017, Ross is the 38th Secretary of Commerce under the US President's Administration.
12 The treatment for removing impurities in metal.
13 A notoriously polluted neighborhood in Niagara Falls, New York.
14 The global courier service.
15 Athletic apparel company that is based in Baltimore.

References

Green, Hardy. *The Company Town: The Industrial Edens and Satanic Mills that Shaped the American Economy*. New York: Basic Books, 2010.

Hall, Elmer J. *Diary of a Mill Town: Recollections of the Bungalows and Sparrows Point, Maryland*. self-published, revised edition, 2004.

Reutter, Mark. *Sparrows Point: Making Steel—the Rise and Ruin of American Industrial Might*. New York: Summit Books, 1988.

Rudacille, Deborah. *Roots of Steel: Boom and Bust in an American Mill Town*. New York: Pantheon, 2010.

4 Looking Beyond the Image of "the Wasteland"

Newark, New Jersey

Lori M. Gibbs

During the 1960s and 1970s, violence, environmental toxicity, cultural decline and physical decay colored Newark, New Jersey's reputation in journalism, art and popular culture alike. The idea of "the wasteland," in physical and cultural dimensions, came to characterize Newark's post–World War II image. Economic conditions in Newark, and the sprawling built environment, mirrored each other in a process of "entropy" described by artist Robert Smithson in the late-1960s. Downtown Newark lost diverse immigrant communities, while the city's physical bounds were reconfigured, emptied and displaced in numerous ways. "White flight" from the city center to the suburbs segregated communities, similar to other American cities at the time. African Americans looking for work found this post-industrial city lacking jobs and support for those in poverty. The development of highways and railways connecting the downtown to satellite suburban developments, and Federal Housing Policies supporting urban renewal, contributed to this complex and increasingly dispirited situation. The role of the local government, and its actions at this time, lead to the general public's frustration with the city's management.[1]

The race riots of 1967 marked an apex of violence on Newark's streets. These riots not only signified a breakdown in civility, which manifested citizen's frustrations with poor-quality housing and health care, poverty and civil rights issues, but also resulted in physical destruction of Main Street.[2] While the city's boundaries and urban qualities reconfigured, the port and maritime activities on the Passaic River split from quotidian life and civic governance. Newark Liberty International Airport and Port Newark (the first shipping container port in the world) significantly refashioned the relation of industry to the city too.[3] These global industries lacked the interest, engagement and immediate connections to the city's population, a quality past industries had fostered in Newark.[4] Newark found itself in a state of crisis.

To challenge the looming image of the wasteland, this study recalls and observes a few (of the many) forgotten places which endure and run counter to the image of Newark as a wasteland, and asks several questions: in a place harboring so much natural, economic, and urban degradation, how have specific architectural and landscape projects remained enduring places for cultural life? If so, which of these places become visible to those willing to set aside the turmoil, the sensational and the shocking? The following examples of apartment complexes, parks and neighborhood streets set out to discover pockets of vibrancy since the dramatic post-industrial downturns of the 1960s–1970s.

Fixing the City

Before the depopulation and dispersal of residents from Newark, it was a bustling 19th-century center of industrial production and cultural activity. Between 1830 and 1930 "Newark's name was carried all over the world on spools of Clark thread, Wiss scissors, Ballantine beer and ale, and 10,000 other locally made products."[5] The dynamism of the city was seen in the establishment of important cultural institutions at this time, such as the Newark Museum in 1909, and the Newark Public Library in 1901. Additionally there were many theaters, department stores and restaurants serving residents and visitors from Northern New Jersey.[6]

In 1921, the New York and New Jersey Port Authority was formed to oversee the port. During the late 1950s major industrial activity in Newark shifted, from the city's center to the port and airport. Port Newark, once a marshland, became the world's first shipping container port, after dredging reshaped the water's edge. In a broader trend of deindustrialization, many factories shut down and heavy manufacturing disappeared from Newark. The focus on international trade at the port separated the water's edge, the port's activities and the surrounding waterways from other uses. Many local industries that benefited from the river as a resource, closed, thus no longer contributed to the vibrant life of Newark as a civic center in the same ways that earlier industry had.[7]

In addition to the closing of factories, the urban core lost even more residents after violence peaked in the riots of 1967. Life magazine featured this "war zone" on its July 1967 cover. The dissolution of civil order was featured in a photograph of a bleeding young boy lying on the pavement—a symbolic description of the dying city. The violence in Newark erupted over police brutality protests, which deteriorated into looting stores; reporters and residents witnessed the killings of 26 people while the National Guard was called in to address the situation.[8] The complex dimensions of civic disorder, environmental toxicity and destruction of the urban core have had lasting effects on the reputation and representations of Newark since these poignant events in the 1960s and 1970s.

Civic activists and government officials vigorously debated the drastic problems of post-war American cities in the 1960s. The debate, about urban renewal projects and highway development in American cities, between Jane Jacobs and Robert Moses in New York City has many parallels with Newark's conditions. *The Death and Life of Great American Cities*, a famous book by Jane Jacobs, a community organizer living in Greenwich Village, was a constructive criticism of "the city." Instead of illustrating the city with images for the reader, Jacobs directs the reader toward the city itself, "For illustrations, please look closely at real cities. While you are looking, you might as well also listen, linger and think about what you see."[10] Her prompt, to look beyond curated images of the city and the prevailing assumptions such descriptions contain, is a fitting method to look beyond Newark's prevailing wasteland image.

The Wasteland

The bleakness in T.S. Eliot's 1922 poem "The Waste Land" describes a world strikingly similar to Robert Smithson's 1967 depiction of suburban New Jersey in *A Tour of the Monuments of Passaic*, New Jersey. T.S. Eliot's landscape is grim and filled with rubbish, a place where depictions of natural features, such as trees, rocks and parched landscapes provide no relief from sun's harsh rays. The ubiquity of such ambient conditions—a natural place with no fecund qualities—parallels Smithson's observations of New Jersey in the fall of 1967, just several months after the devastating riots in Newark's downtown.[11] This account describes a post-industrial landscape surreal in its dystopian qualities on the fringes of New York City. Smithson's finds New Jersey filled with monuments to industry and highway construction—activities shifting the terrain's features, culture and eroding the boundaries of the city.

Using his instamatic Kodak camera, Smithson visualizes his description of this "wasteland," categorizing it as: colorlessness, yet illuminated by a "monstrous light-bulb" sun, filled with "holes" and lacking a center. Smithson is fascinated with industry's impacts upon the landscape, sprawling and "bland" suburban life "littered with shiny cars," a life that lacks a "rational past." In short, Smithson is witness to the dispersal of civic life from Newark's urban core into pockets of suburban development, partly made possible by new roads, (and the growing import of foreign cars at Port Newark).

Smithson sees this new type of urban formation as a wasteland in cultural terms too, "that unimaginable suburb could have been a clumsy eternity, a cheap copy of The City of the Immortals."[12] The isolation and decentralization of everyday life into single-family homes and sidewalk-lined streets create vastly different, highly manicured settings that prioritize the automobile over vibrant pedestrian filled streets. In many ways, this portrayal of New Jersey's cultural and physical constitution has not changed much in the years since.[13]

Exploring New Jersey became fodder for Smithson's imagination and material for his Earthworks art. Smithson's Earthworks projects display heaps of raw geological materials collected from New Jersey, such as limestone and slate. These materials are displayed in an in-between state—not quite "natural" anymore, and yet not fully formed by industrial processes into usable objects, either. Besides Smithson's critique of the New York art-world, these Earthworks also express a metaphor about the changing landscape, shifting urban conditions, culture and post-industrial configuration of New Jersey.

In this characterization of the wasteland, dirt represents a ground in flux, a natural substance harboring opposing potentials: cultivation and decomposition. Dirt either roots growth, or acts as sediment burying the expired. This twofold quality creates a moment of ambiguity, which Smithson observes and counters; he questions the paradigm of technological innovation following a ceaseless pattern of progressive linear development.

Smithson comments on the destruction inherent in the "rebuilding" of a suburban ideal, "With such equipment construction takes on the look of destruction; perhaps that's why certain architects hate bulldozers and steam shovels. They seem to turn the terrain into unfinished cities of organized wreckage."[14] Counter to the word's etymology, construction has become a decomposing force, creating a chaotic state of decay that ultimately lacks the promise of completeness, organization and vibrancy. This provisional state of the landscape and development, not quite finished or discarded from previous use, is found in a state of dissolution or, the natural state of "entropy." The concept of entropy addresses an understanding of nature and natural landscapes as embedded within "a process of ongoing relationships existing in a physical region" rather than conceptualizing nature as absolute, fixed and highly formalized. For Smithson, nature and industry's temporal qualities parallel the city, and its culture, in a state of disintegration.

The "spoilt" quality of New Jersey, and Newark, is predicated upon a deeper belief in the purity and moral authority of "nature" as a categorical distinction, separate and pure, from culture and the constructions of man.[15] At this precise moment, Smithson evokes a particular distinction between the "natural" and his discontents with suburbanization taking shape—eating away at both natural landscapes of the river and dissolving the vibrancy of the city's urban core. The city and Nature (as a raw categorical entity and viewed in opposition to the city) are both changing, due to highway construction and deindustrialization, and are both qualified by the natural laws of "entropy" Smithson sees in this setting.

The qualities of incompleteness Smithson finds along the Passaic River could be considered meditations on the aesthetic category of the ugly. Related to the category of beauty, the ugly represents the "the object that has gone wrong" and as the "privation of what should have been." A state of "ugliness" complicates beauty, and "achieves an ambiguous status" due to its incompleteness. The ugly represents a state of contamination, deformation or displacement.[16] These qualities closely relate to the idea of the wasteland and the setting Smithson observes as in flux, filled with traces of unrealized futures he categorizes as monuments to industry. New Jersey's landscape is filled with out-of-place debris, lacking the solidity of the organized, complete and rational city. Marks of 19th-century industry on the landscape cast a shadow over contemporary life and contaminate the naturalness of the soil and existing waterways; in this sense Newark is a place far from the ideal completeness and dynamism of a vibrant urban center, or from harboring nature's purity.

The "Natural" Wasteland

Yet, even before the "contaminants of civilization" spoilt New Jersey, the "natural" marshes and swamp, were already viewed as a wasteland. White cedar forests covered the marshes and swampland, prior to the detritus of industrial pollution filling the Passaic and Hackensack Rivers. Deemed unsuitable and untenable for development, measures of active filling and draining turned the landscape into a productive grain yielding meadow. Now called the Meadowlands, yet no longer agriculturally productive, this area has a long and complicated history of ownership and development beyond these initial transformations by Dutch settlers.[17]

The land's "natural" state was framed unpleasantly in the 1960s as "Famous as a breeding ground for mosquitos, a run for muskrats and a place to dump that which is no longer useful."[18] Its "natural" qualities were not viewed as immediately beneficial for any usable activity—besides dumping refuse, and functioning as a receptacle for the unsightly or unwanted. In 1962, the meadowlands were described as 25,000 acres of the "most valuable unused real estate in the U.S.A" yet a problem because it was a "swamp" in the middle of an urban complex. The view of the area as a precious place with an important natural ecological balance had not yet come to fruition, (despite an evolving environmentalism discourse at this time). This area remained unsuitable for expanding the city, due to the soft and unstable soil of the landscape. This view coincided with landfilling to stabilize the soil in hopes of development. The category of "wasteland" slipped between describing the "natural" state of a landscape incapable of grounding urban development, to the post-industrial city in ruins; the category of the ugly shifted from the natural to the man-made. The Meadowlands was a place to dump anything and everything unwanted in New York and Newark; including the debris of New York's Pennsylvania Station (demolished in 1963), and the haphazard disposal of toxic materials on land owned by Newark's Archdiocese.[19]

Envisioning what the "natural" can be has clearly impacted the Meadowland's treatment as a no man's land—a place for highways, railways and dumping (both legal and illegal). Equally the city's buildings and parks are framed and informed by dreams of rebuilding the swamp—some endure despite the challenges of post-industrial dynamics.

Repopulating the City in the Late 1950s–Early 1960s

Ideal models for urban life are often posed as static, unchanging solutions, that once realized, will comprehensively address all future needs and activities. However this investigation of Newark highlights that while some forms of urban development endure and/or adapt to drastic reconfigurations of society, others become untenable and disintegrate.

The loss of inhabitants deeply affected Newark, decline started as early as the 1940s;[20] residents left for new homes in the suburbs, benefitting from the Federal Housing Authority's policies outlined in 1949. These measures "ghettoized" American cities like Newark by "subsidizing the flight of middle class workers to the suburbs."[21] These policies, as well as a culture of corruption and political fragmentation, are seen as the reasons "Why Newark Struggles."[22] The Federal Housing Act of 1949 is largely thought to have segregated white and black communities. Title 1 of the Federal Housing Act provided money to clear slums in cities and financed new developments to be built in their place. Other aspects of the Federal Housing Act subsidized mortgages for whites, and blatantly excluded black communities through stipulations in mortgage lending criteria. Therefore only some benefited from moving to new suburban communities outside of the city's traditional boundaries.

However, not all the federal aid went to build suburbia or unsuccessful housing projects. The Pavilion and Colonnade Apartments in Newark's North Ward represent a counter example, intending to repopulate Newark in the late-1950s. This housing complex was the first urban renewal project in Newark funded with Title 1 funds. Just seven years prior to Smithson's exploratory visits to the Garden State, modernist architect Mies van der Rohe designed a series of three apartment towers situated in an open space in Newark's North Ward, at the south end of Branch Brook Park. This project aimed to attract middle-income residents to the city. Mies van der Rohe, a former student of the Bauhaus and German émigré to the US in the 1930s, partnered with Herbert Greenwald, a "philosopher turned developer."[23] The architect and developer created projects in Chicago and then expanded to Newark. The Newark development sought to add to the "continuous city landscape" rather than present an absolute or totalizing vision for the city. Greenwald "believed in the city. He had a relentless vision that demanded open spaces, the subordination of the automobile, and the nourishment of human values."[24] Mies also shared a view of betterment for the city, which he realized architecturally, project-by-project.

Mies' design features steel and glass towers rising from an uninterrupted ground plane—an intended corrective to the squalor of the crowded city and the dysfunction of the suburb (as Title 1 projects intended to ameliorate unsanitary conditions of the slums). The glazed high-rises were built with anodized aluminum curtain walls, intent on "expressing the structure of the emerging, decentralized urban field one project at a time."[25] This vision was not a totalizing strategy for the city, but rather one "intended to control urban decentralization" and to create a place that is "noble and uplifting, quiet and calm, vital and monumental."[26] This localized project not only set out to combat the ills of the pre-war city, and correct its spatial qualities fostering unhealthy conditions, but also counter those of suburban life in the post-war era by drawing people back into a livable Newark and into an urban setting of higher population density. Within a short walking distance to Newark's downtown the housing development's residents could live suspended within a floor-to-ceiling glass bounded space, opening one's field of vision to the surrounding city and views of Manhattan beyond. The architect intended to "achieve a sense of openness and elevation in order to give each resident a feeling of personal freedom even though he is close to the heart of the city."[27] This organization of space countered the dark, crowded conditions associated with the slums; and is distinct from Olmsted's parks which immerse the individual within a representation of natural landscapes. Here the resident is immersed in the view of the city itself, yet given a private, tranquil space within which to contemplate this surrounding environment (Figure 4.1).

Mies and Greenwald's shared vision is further explained by architectural historian Detlef Mertins, as an architecture that would: "use technology and rationalization to redeem the soullessness of American capitalism, mass society and mediate culture and would enable freer ways of living within it."[28] The developments were financially successful too. In fact Greenwald's developments were cheaper than his competing developers; partly due to the low fee Mies charged. These intentions also promoted the value of aesthetics in everyday life and the idea of open yet ordered space, compacting the building's footprint and providing a pedestrianized ground level with no obstructions, not even apartment units.

The pedestrianized space and largely uninterrupted ground plane stand in opposition to the many principles Jane Jacobs outlines for cities, however this development has endured and remains an affordable home to many residents today. Recently, residents protested against the building owner's proposal to erect a six-foot security fence around the perimeter of the development. "Commission members unanimously

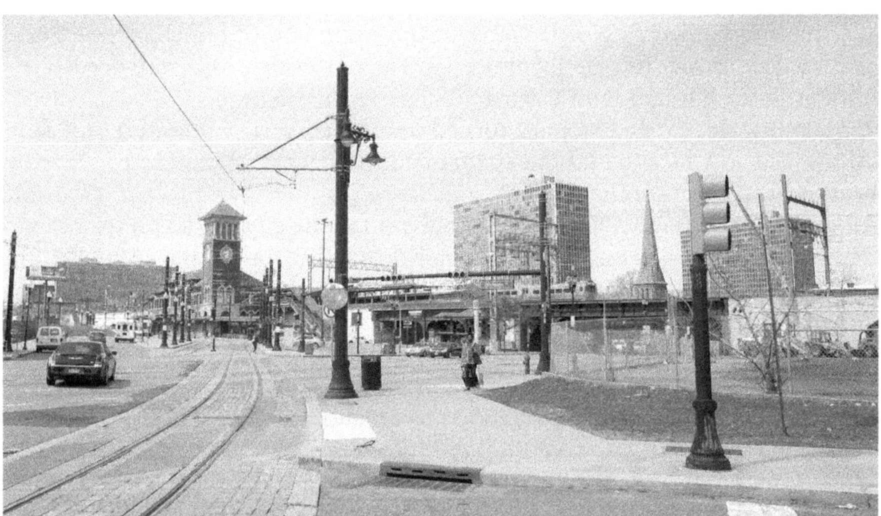

Figure 4.1 Lori Gibbs, *Broad Street Railway Station and Pavilion Apartment Towers*, in the distance, 2017

rejected the plan because it wasn't in keeping with the aesthetics that van der Rohe created for the Colonnade and the Pavilion."[29] The ideal, of open yet highly ordered space for pedestrian leisure on the ground level, remains a priority for residents keenly aware of the qualities making this development livable—including the importance of aesthetics. In this sense, the Pavilion and Colonnade development remains an enduring setting in Newark, representing spatial ideas intending to repopulate Newark in the early post-war decades.

Other Cities-Within-a-City

The Mies complex has outlasted other failing redevelopments constructed in Newark at this time, most notably the Christopher Columbus Housing Projects—once sandwiched in between the three Mies towers. Built in the 1950s, four of these public housing towers were imploded in 1994. This literal destruction represented and manifested the failures of federal high-rise public housing development programs initiated in the 1950s. Mayor John Sharpe is quoted in the *New York Times* stating "This is the end of an American dream that failed." His words came at the ceremony just before one of the largest demolitions of American public housing took place.[30] Clearly, not all design intentions and urban plans are capable of enduring; the lack of maintenance by management of these homes added to the conditions described, by senior Federal housing official James E. Baugh in the 1980s after touring the complex, as a place where "No human should ever have to live . . . no animal should ever have to live." This dynamic also factored into the deterioration of the hope that high-density housing could be a replacement for the pre-existing "slums" as an acceptable model and reality for public housing, in this context.[31] The disintegration of such a model requires a future remaking, and parallels Smithson's entropy metaphor for New Jersey.

In 1977, the *Star Ledger* published a full-page mosaic of images with the headline "A message for those who think Newark died 10 years ago". The images showcase 17 urban renewal projects built after the infamous riots in 1967. The page is populated with buildings and complexes representing "tangible steps forward" which "stand as dramatic testimony to our city's spirit and its superior location." This advertisement, sponsored by the Greater Newark Chamber of Commerce and Newark Redevelopment and Housing Authority, clearly tried to reshape the image of Newark. By showcasing the progress made in literally reconstructing the city, evidence is laid out to challenge the image of the "wasteland" with buildings that symbolize a "promise of a new and great era for the Newark of tomorrow."

The city depicted here is now home to a Serta Mattress Plant, the Ideal Toy Corporation Plant, Essex County Community College, N.J. Institute of Technology, several hospital complexes and other college developments, a new passenger terminal at Newark International Airport, a jail and county courthouse, as well as 6,000 housing units. As convincing as the spread of images tries to be, the ambiance of the city and the daily activities of its residents are largely absent from this description. Oblique and aerial views showcase the various building entrances and grandeur, but fail to describe the street life or the interconnections between these large developments within the city. The ambitions advertised here revolve around the idea that renewal of the city lies in the singular dimension of constructing large, bold physical interventions, and lacks a focus upon interlinking such interventions to other locations and cultures within Newark's wider terrain.[32]

The Gateway Plaza Complex is one featured development in the heart of downtown Newark, constructed just shortly after the 1967 riots (Figure 4.2). A brochure for the

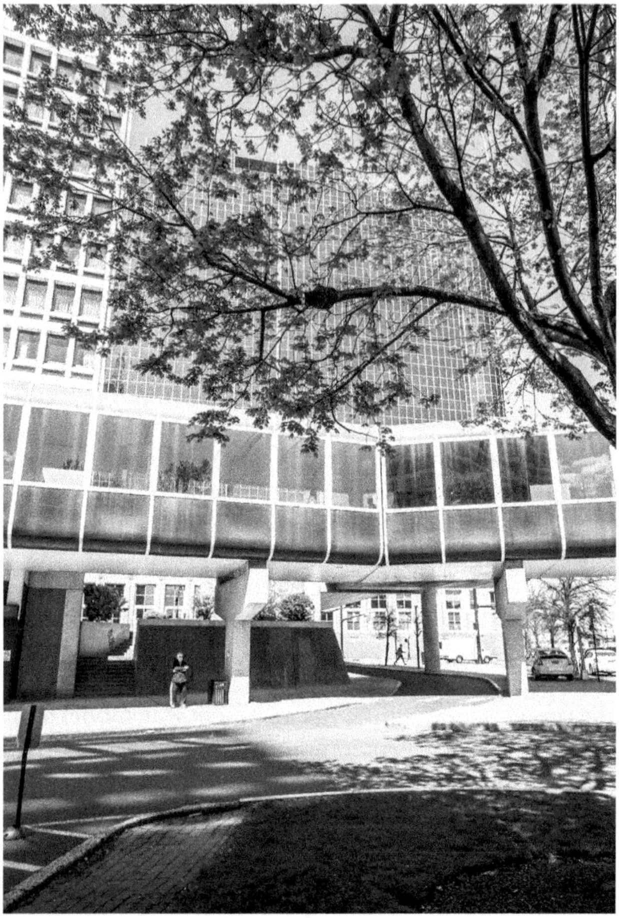

Figure 4.2 Lori Gibbs, *Gateway Plaza Pedestrian Bridge*, 2017

Gateway Complex advertises a sumptuous, luxurious, secure, vast city-within-a-city: "America's most important office, shopping and entertainment complexes." Its convenient location, and interconnection to Newark Penn Station, even boasts the complex as a seven-minute helicopter ride away from New York City. This type of luxury almost seems unthinkable, in opposition to the intense suffering experienced by the community during the riots. This development internalizes public space and creates a barrier between those using the complex, and the broader dimensions and uses of public space. Additionally it excludes any representations of the natural, through landscape or any other outdoor features.

White-collar workers fill the internalized streets of the complex during the day, but after the workday and on weekends there is an eerie silence. One can walk two city blocks without ever leaving the complex. The irony of the "city-within-a-city" design as an urban strategy, is that this microcosm actually opposes generating vibrant street life; perhaps that was a goal after the riots of Newark destroyed many businesses. The internal passageways that replace ground level streets are empty due to the lack of

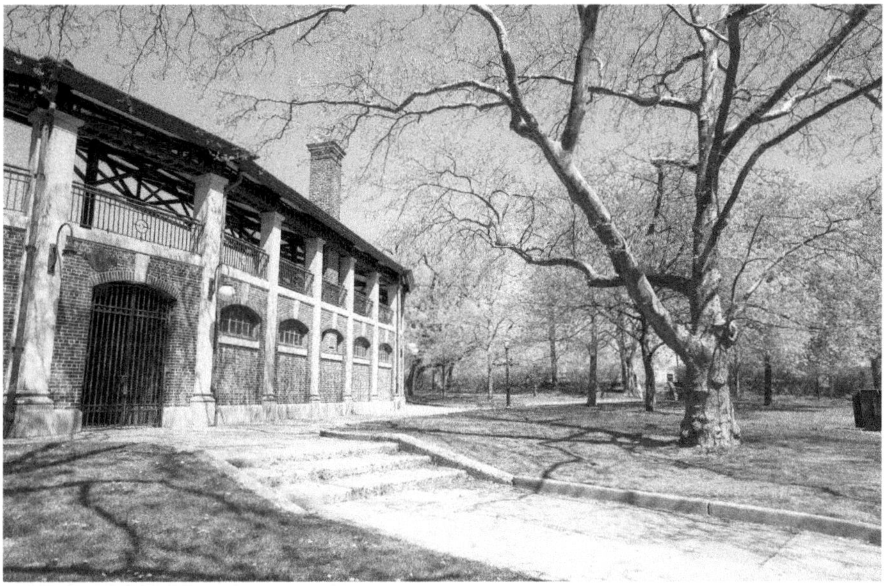

Figure 4.3 Lori Gibbs, *Riverbank Park*, 2017

variety of occupants and activity once business hours are over. This type of complex largely protects white-collar workers from the riff-raff of Newark's streets; further splitting communities within the city itself, and fails to reintegrate different residents within urban spaces. Supposedly envisioned as an interior "main street" Gateway fails to truly replicate the vibrancy of traditionally successful commercial corridors, while also excluding elements of the natural in its hermetic containment of commercial and office spaces that adjoin to large multi-story parking garages.

Just outside the station's south door is Ferry Street (an almost) prototypical "high street" lined with small businesses, shops, restaurants, fish and vegetable markets, bars and cafes. These businesses cater to low-to-middle income residents; many are Portuguese, Spanish, Brazilian, Ecuadorian and Mexican immigrants. During the day Ferry Street bustles with diverse activity, accommodating football (American soccer) celebrations, seasonal festivals, displays holiday decorations and creates a safe space for Halloween trick-or-treating. In this sense, it hosts much more than the internal walkways of Gateway Plaza ever could—and yet these two places exist side-by-side. In Newark's 2012 master plan images of Ferry Street and the Ironbound neighborhood are models for replication across the city—not images and qualities of Gateway Plaza. Yet, despite the close-knit vibrancy of the Ironbound community, it has faced sordid threats from the industry that once functioned there.

Environmental Activism and Community Organization

Historically, the Ironbound neighborhood, just south of Newark's Penn Station, has been a center for industrial activity along the banks of the Passaic River. Perhaps completely interiorizing urban space at Gateway Plaza was partially a response to the residues of pollution and the environmental contamination problems peaking in the

1970s from factories and toxic industry in the immediately adjacent Ironbound. In 1978, the Ironbound Community Corporation formed, in protest of environmental contamination and other threats to the health of the community. "Ironbound," the name of the neighborhood, is itself rooted in industry. It literally describes the geography of the neighborhood as bound by railroad tracks. Now, railroad tracks, the Passaic River (a Superfund site), and highways define the neighborhood's edges.

In a small corner of the local library on Van Buren Street, a modest desk, file cabinet, sign-in notebook and large folio make up the Ironbound Environmental Justice History and Resource Center. The trilingual publication of Ironbound Voices embodies the diversity and unity of the Portuguese, Spanish and English speaking community, representing the vibrancy of the resident population. Among young elementary school kids completing homework at other tables nearby, one can sift through the clippings about DDT production, and Agent Orange (1966–1968) leaving unhealthy amounts of dioxin present in the neighborhood. Residents found the state and federal government had knowledge about the levels of dioxin found, thought to be higher than anywhere else in the US.[33] Here the dust and soil in the area carry agents of poisonous decay.

An independent commission was formed, and tested other areas of the neighborhood to see if spreading of dioxin had occurred, and it had. The Diamond Alkali Superfund site along the Passaic River is now capped with concrete to contain the dioxin (which cannot be easily disposed). Councilman Amador discussed the fate the Passaic River, supporting it, "as a natural entity that should be treated as the heart of the community."[34] Neighborhood activists have spent years fighting for toxins to be properly handled and contained to prevent harm to human health and to ensure the 50,000 residents are not further harmed by past industrial activity; not to mention the current impacts of the airport and contaminants from trucking along highways servicing goods from the port to outlying areas.

The Ironbound is also home to Riverbank Park, one of the earliest Essex County parks designed by Frederick Law Olmsted in Newark, 1908 (Figure 4.3). In 1867, citizens had the ambition to build the Nation's first county park system and involved Olmsted. Credited as a founder of landscape architecture in the United States, Olmsted designed hundreds of parks across the country (Central Park is one of the most iconic). Newark parks are visionary works of landscape architecture, which in Olmsted's eyes were "designed, with the intention of producing, through the exercise of the natural perception, a certain effect upon the mind and character of those who approach it."[35] The impetus for parks sought to bring "means of healthful enjoyment to every class of our citizens that may feel disposed to partake of its benefits."[36] At the time of Newark's industrial prosperity, Riverbank Park was more easily accessible to factory workers in the East Ward, than the larger Branch Brook Park. This idea of the natural, as a setting of vegetation and topography, is designed in ways less perceptible to the uninformed eye. In this sense it resists Smithson's fascination for the uncertainties and ambiguities of disintegration; as Olmsted's designs alter given conditions, and account for continual regeneration of vegetation and changes of seasons.

The intricacy of Olmsted's design for this park is remote from locals' everyday use. Yet, the concerns Olmsted's practice had with the representation of nature, embodied by the landscape of the park, persists in its everyday recreational uses, and serves as a setting for athletic training, sporting events and playgrounds. Olmsted's parks endure beyond their representational qualities of nature, in the diverse use of plantings

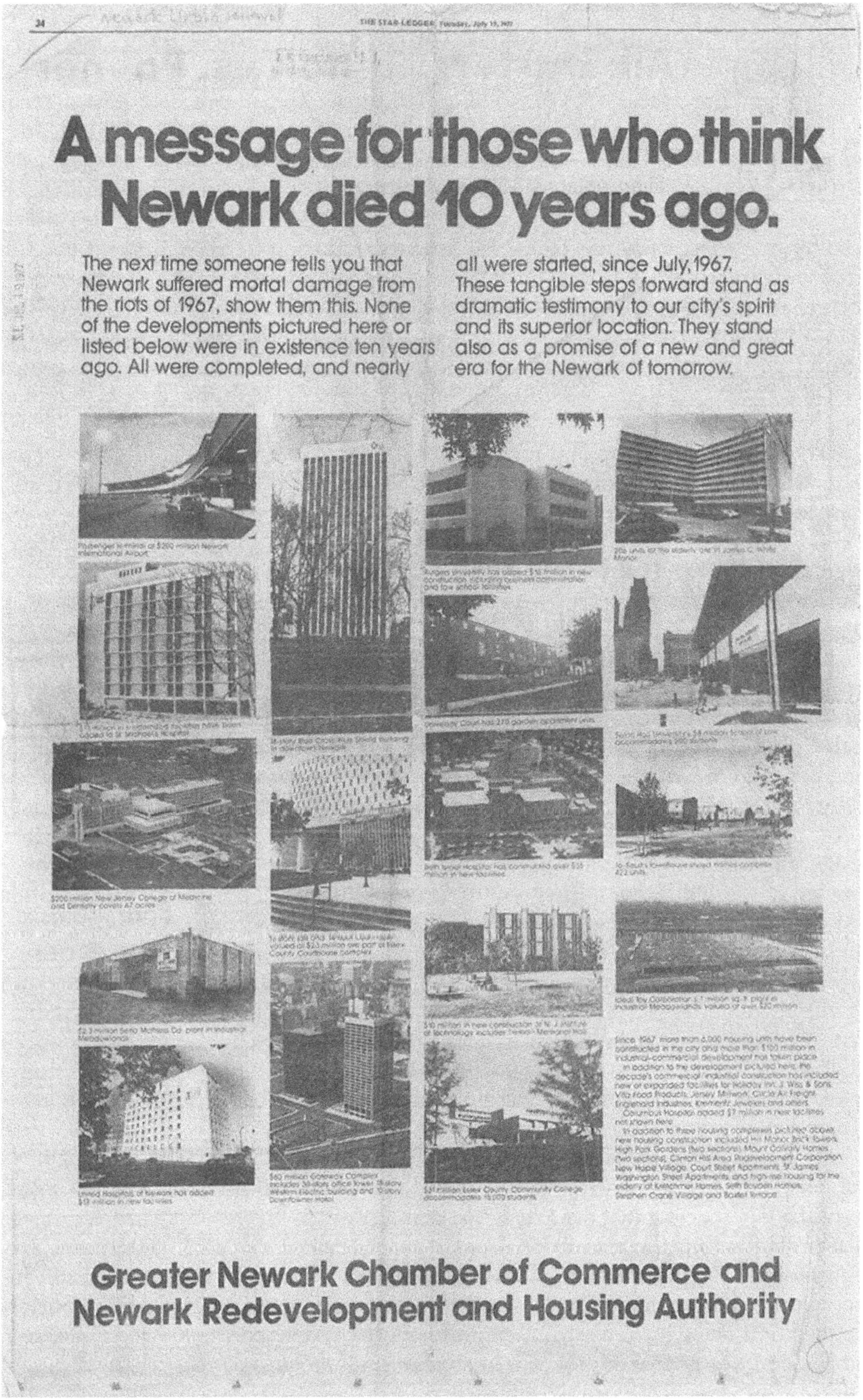

Figure 4.4 *Star Ledger* Advertisement for Urban Renewal Projects in Newark, July 19, 1977

and bold forms that contrast the city's concrete structures, industrial complexes and asphalt pavements.

To portray the "natural," Olmsted's practice used large open fields, arrangements of groves, trees, plantings and water features to craft a space separate from cars and the grit of the city. These works of landscape architecture are successful because of their representational dimensions of "nature" that seamlessly dovetail with the physical qualities of foliage, running water and embedded architectural features. Up keeping this representation of nature reveals that, despite their appearance, Olmsted's parks are not "natural" in themselves. Thoughtfully designed spaces for the purpose of recreation and providing a space of leisure for those living in industrialized urban places, perhaps would not be quite what they are without other countering dimensions of urban life (work, industry, compact living spaces, high density). In Smithson's words, "Olmsted made ponds, he didn't just conceptualize about them."[37]

Branch Brook Park, the first manifestation of this vision of civic life "in the park" for Newark also still exists; Mies's residential development lines its southern edge.[38] This park is 360 acres and home to more than 2,000 cherry blossom trees in Newark's North Ward, more than the promenade in Washington DC.[39] These trees were originally a gift from Caroline Bamberger Fuld in 1928, a relative of the once-famous Bamberger department store in downtown Newark. The planting of the trees took four years to implement, and the 50-year life span of the original trees has already been surpassed; few of the original plantings are left. In 1974 community members formed Friends of Branch Brook Park and in 1976 the Newark Cherry Blossom festival was created. These, among other activities, oversee the continued maintenance of both a physical setting and a powerful idea for the role of nature in the industrial city, long beyond Olmsted's original intentions and constructions, and with the capacity to endure the drastic changes in Newark.

Although Olmsted's parks present the image of a natural landscape through their design, this area was reclaimed from a pre-existing swamp. The realization of the park entailed re-working the landscape, cutting and filling, demolition of existing buildings, and carefully planned plantings to form this "natural" setting. As Smithson points out, "Olmsted's parks exist before they are finished, which means in fact they are never finished; they remain carriers of the unexpected and of contradiction on all levels of human activity."[40] Designing a specific idea about nature's role in the industrial city was key to establishing park spaces in Newark. The parks were not simply created by cordoning off an unused parcel of land, and these parks demand continuous upkeep. Perhaps, this can be seen as a means of design considering and engaging with dynamic natural processes of change over time.

This particular idea of nature in the city endures in Newark, as it does in Central Park and Prospect Park (other Olmsted creations). In 1996 Ironbound's SPARKS organization successfully thwarted the destruction of Riverbank Park for development, maintaining this recreational oasis for residents. The park closed in 2001 for remediation, as the contractor Marsellis-Warner Corp used contaminated soil in the renovation—containing PCBs, arsenic and other toxins.[41] This hazard has since been rectified and the park remains open and well used. The park, as a representation of nature, endures primarily because of the continuous engagement of the community. After a century since its initial construction, residents have continually maintained and renovated the park, long after its initial plantings have blossomed. The capacity for the park to accommodate changes in the vegetation over time is part of its design. The representation of nature, in this case, also prompted the remediation of the soil from industrial contamination; allowing this part of the Passaic River's banks to be inhabited once again.

Conclusions: Enduring Places and Representations

As a measure of resistance against the dissolution of city's vibrancy in the early 1960s, Mies' creation of open space and views of the surrounding urbanity were an effort to re-engage and reorganize residential life in Newark's urban core. This did not rely upon creating a "natural" setting, but rather a highly ordered yet open space privileging views out over the city. Frederick Law Olmsted's aesthetic interventions created a microcosm of vegetation and topography that represented nature in the industrial city—spaces that function as places for therapeutic repose from the harshness of factory work, or today's office work. Though drastically different from Mies' apartment complex, these places coexist on the southern end of Olmsted's Branch Brook Park, and have endured despite drastic changes in Newark's physical, social and cultural constitution. These visionary designers are not often associated with the city of Newark, yet these places are windows into looking beyond the descriptions of Newark as a wasteland. These projects, Riverbank Park, and Ferry Street, begin to reveal an urban texture filled with places continuing to foster ideas about the role of nature and spatial organization in the city, despite the struggles Newark has faced.

In the case of Gateway Plaza, internalized public space shuts out the unsightly aspects of Newark, as well as references to nature in the city. City-within-a-city developments, like Gateway Plaza, continue to be adapted and inserted into Newark, despite the limitations of such "superblocks." Richard Meier, a world-renown architect from Newark, just completed Teacher's Village, a development of several blocks of subsidized housing for teachers and charter schools. Perhaps this will become another method for repopulating the city, if it is successful.[42] However, one of the dangers may be the enclosing microcosm of a "city-within-a-city" typology, splitting specific communities away from each other. Yet, the scale and street level engagement of Meier's development may hold much more sensitive architectural qualities to reform downtown Newark, by not excluding those it does not immediately serve, and reinserting residential units back into the downtown.

As of 2017, a speculative riverfront development is planned to bring a new formulation of the "natural" to the Passaic River. Designs proposed by James Corner Field Operations (the same landscape architects of New York's Highline) will revive the banks of the Passaic, and represent a return to valuing "nature," and its image as a setting for Newark residents. This is an interesting turn, taking a superfund site and remediating its "natural" elements (soil and dead riverbed) into a fecund site for both the representation of nature in the city but also giving form to new settings for a repopulated Newark. Rekindling the representation of nature in the city, as a focal point for residents, is a topic under active consideration again. Despite the looming descriptions of Newark as a wasteland, and the harsh realities of its environmental hazards, poverty and violence, the city remains a place populated with enduring settings and communities where not all was lost; it just might not appear to outsiders as "spotless" as neighboring New York. Investigating Newark on its own terms allows one to look beyond the image of the wasteland, and learn from the attitudes, ideas and designs that continue to impact and reconfigure this particular cityscape.

Notes

1 For a more detailed and nuanced discussion of Newark's political environment in the 1960s, and discussion of Louis Danzig's unchecked power to reshape the city as the head of the Newark Housing Authority, please see: "Planning, Slum Clearance and the Road to Crisis in Newark" by David Levitus, online at The Newark Metro website: www.newarkmetro.rutgers.edu/reports/display.php?id=173
2 Oral histories from residents of Newark at this time are discussed in relation to the broader historical, social and political context of the city in: "Silent No Longer: Voices of the 1967 Newark Race Riots" by Kimberly B. Siegal (University of Pennsylvania, senior thesis in Honors American History, March 2006). Accessible from University of Pennsylvania Scholarly Commons.
3 See the Port Authority's website for a brief history of the development of containerization in the shipping industry at Port Newark and in Elizabeth, New Jersey: www.panynj.gov/port/history.html
4 The many donations of fine art and sculpture housed in Newark's Museum are evidence of such contributions.
5 *Knowing Newark*, "Town of Cottage Industries Becomes a Manufacturing Empire," 22.
6 See *Knowing Newark* for various articles describing Newark's cultural scene at this time.
7 Today, the natural features of the city's geography—the river's banks—mostly lie fallow and fenced off from public use or paved as parking lots for commuters.
8 *Life* (July 1967).
9 Curiously, this debate took place a mere eight miles from Newark, however Newark is not mentioned in the many examples Jacobs utilizes to explain her principles for better cities.
10 Ibid.
11 See Robert Smithson, and Jack Flam (ed.), *Robert Smithson: The Collected Writings* (Berkeley: University of California Press, 1996).
12 Smithson and Flam (ed.), *Robert Smithson*, 56.
13 Philip Roth uses Newark as a setting for his novels, such as *American Pastoral.* Also, the Sopranos, an HBO fictional series, utilizes Newark and Elizabeth, New Jersey as a backdrop integral to the show's narrative. The portrayal of New Jersey reinforces the moral corruption of organized crime families in the late 1990s–early 2000s. The Sopranos are also in the "waste management" business.
14 Robert Smithson, "The Sedimentation of the Mind: Earth Projects (1968)," in *Robert Smithson, Collected Writings*, 101.
15 Lorraine Daston and Fernando Vidal (eds.), *The Moral Authority of Nature* (Chicago: University of Chicago Press, 2004), outlines an in-depth historical and theoretical treatment of this topic.
16 Cousins, Mark. "The Ugly," in *AA Files*, No. 28 (Autumn 1994), Architectural Association School of Architecture, 61–64. Cousins, situates his study of the ugly in psychoanalysis, describing the reaction of one who "turns away" or "turns their nose up at" whatever exhibits qualities of the "ugly"; this visceral reaction to Newark is still commonplace within the local context.
17 For an in-depth history about the Meadowlands see: Robert Ceberio, and Ron Kase. *New Jersey Meadowlands: A History* (The History Press, 2015).
18 *Newark Commerce*, VII, 5 (November 1962).
19 See Ceberio, and Kase. *New Jersey Meadowlands: A History.*
20 "Newark the Living Downtown Plan," 2008; http://planning.ci.newark.nj.us/wp-content/uploads/2014/09/econ_redev_plan_Downtown.pdf
21 Newark Mayor Kenneth A. Gibson, "10,000 Home Community in Heart of Newark," *The Star Ledger* (Thurs. November 18, 1971).
22 See Robert Curvin's book for a detailed and more nuanced discussion: *Inside Newark: Decline, Rebellion, and the Search for Transformation* (Rutgers University Press, 2014), 16.

23 Berger, Miles L. *They Built Chicago: Entrepreneurs Who Shaped a Great City's Architecture* (Bonus Books, 1992), 237.
24 Ibid., 238.
25 Mertins, Detlef. *Mies* (Phaidon Press, 2014); 402.
26 Ibid.
27 *Newark Evening News* (August 25, 1959).
28 Mertins, *Mies*, 314.
29 Carter, Barry. "Newark's Colonnade Residents Don't Want to Be Fenced in," *The Star Ledger* (November 9, 2016).
30 Levy, Clifford. "4 High-Rises Torn Down by Newark," in *The New York Times* (March 7, 1994).
31 However, if one looks at high-rise public housing in other contexts, such as in Hong Kong, the contextual conditions and maintenance of these complexes are far different and continue to be settings that succeed as healthy and safe, even pleasant places to live. Which also helps to shed light on the fact that the physical form or type is not the only component to creating an enduring 'setting.'
32 Curiously, Laurajane Smith discusses the discourses of heritage in her book *Uses of Heritage*, and describes the growing concern with material property and buildings in the 1960s and 1970s, in part due to rebuilding in Europe after the World Wars, but also due to massive urban renewal projects in the United States. The comparison of a movement that cleared away the existing problems, also cleared away street patterns, sidewalks, small businesses, markets and informal relationships between residents, is an almost unexpected bedfellow with the destruction of war. But the razing qualities of both of these types of destruction is clear in the case of Newark, where such rebuilding did decimate communities, local businesses and the immaterial bonds that equally gave vibrancy to Newark.
33 "Dioxin in Ironbound," *Ironbound Voices*, 6, 3 (July 1983)
34 Mansnerus, Laura. "Newark's Toxic Tomb; Six Acres Fouled by Dioxin, Agent Orange's Deadly Byproduct, Reside in the Shadow of an Awakening Downtown," in *The New York Times* (November 8, 1998).
35 Vaux Olmsted and Company, *Report on a Site for a Park at Newark*, 1867, in Kathleen P. Galop and Catharine Longendyck. *Images of America: Branch Brook Park* (Arcadia Publishing, 2007).
36 Galop and Longendyck, *Images of America*, 7.
37 Smithson, 159.
38 For a comprehensive overview of Olmsted Parks in New Jersey see: *Olmsted Parks in New Jersey*.
39 Essex County Park website: www.essexcountyparks.org/parks/branch-brook-park/about
40 Smithson, 160.
41 See Riverbank Park's website, "History" section: http://riverbankpark.org/index.php/faq/22-history
42 This project was partially funded by Goldman Sachs, as a one of the firm's "Impact Investing Projects." It seems now Goldman Sachs is hedging its bets on Newark's renewal.

References

"Redevelopment Program for Detroit." *Architectural Record* 127 (April 1960): 170–173.

"Redevelopment Program for Newark." *Architectural Record* 127 (April 1960): 174–178.

Baum, Kelly (et. al.). *New Jersey as Non-Site*. Princeton University Art Museum, 2014.

Berger, Miles L. *They Built Chicago: Entrepreneurs Who Shaped a Great City's Architecture*. Bonus Books, 1992.

Boettger, Susan. *Earthworks: Art and the Landscapes of the Sixties*. University of California Press, 2004.

Broid, Andre. "3 Glass Towers Rising in Newark." *Newark Evening News*, August 2, 1959.

Carter, Barry. "Newark's Colonnade Residents Don't Want to be Fenced in." *NJ.com*, November 9, 2016.

Ceberio, Robert and Ron Kase. *New Jersey Meadowlands: A History*. The History Press, 2015.

Colquhoun, Alan. *Essays in Architectural Criticism: Modern Architecture and Historical Change*. Oppositions Books & MIT Press, 1981.
Cronon, William (ed.). *Uncommon Ground: Rethinking the Human Place in Nature*. W.W. Norton & Company, 1996.
Cummings, Charles F. *Knowing Newark: Selected Star-Ledger Columns*. The Newark Public Library, 2014.
Curvin, Robert. *Inside Newark: Decline, Rebellion, and the Search for Transformation*. Rutgers University Press, 2014.
Daston, Lorraine, and Fernando Vidal (eds.). *The Moral Authority of Nature*. University of Chicago Press, 2004.
Galop, Kathleen P., and Catharine Longendyck. *Images of America: Branch Brook Park*. Arcadia Publishing, 2007.
Goldman Sachs, "Impact Investing: Teacher's Village Redevelopment Project." retrieved from www.goldmansachs.com/what-we-do/investing-and-lending/impact-investing/case-studies/teachers-village.html.
Ironbound Voices, Ironbound Community Corp. Newark, NJ, 1978–2001.
Kempson, John W. "Return to Cities Trend Is Sparked." *Newark Evening News*, March 5, 1960.
Kolva, Jeanne. *Olmsted Parks in New Jersey*. Schiffer Publishing Ltd., 2012.
Leatherbarrow, David. *Topographical Stories: Studies in Landscape and Architecture*. University of Pennsylvania Press, 2004.
Levy, Clifford. "4 High-Rises Torn Down by Newark." *The New York Times*, March 7, 1994.
Massnerus, Laura. "Newark's Toxic Tomb; Six Acres Fouled by Dioxin, Agent Orange's Deadly Byproduct, Reside in the Shadow of an Awakening Downtown." *The New York Times*, November 8, 1998.
Massola, Jessica. "Designer Behind NYC's High Line Developing N.J. Park Expansion." *NJ.com*, June 15, 2016.
Mertins, Detlef. *Mies*. Phaidon Press, 2014.
"Colonnade Apartments." *Newark Evening News*, August, 25, 1959; September, 27, 1960.
"Half of Colonnade Pk. Tenants from Suburbs." *Newark Evening News*, June 22, 1960.
"Colonnade Ceremony." *Newark Evening News*, November 20, 1961.
"Pavilion West Apts. Ready." *Newark Evening News*, September 9, 1962.
Olmsted, Frederick Law, Charles Beveridge, and Carolyn Hoffman (eds.). *The Papers of Frederick Law Olmsted: Writings on Public Parks, Parkways, and Park Systems*. Johns Hopkins University Press, 1997.
Plath, Karl. "Mies Legacy: How Famed Architect Shaped Chicago With Aid of 3 Developers." *The Chicago Tribune*, May 4, 1986.
Smithson, Robert and Jack Flam(ed.). *Robert Smithson: The Collected Writings*. University of California Press, 1996.
Strunsky, Steve. "Newark Breaks Ground on Riverfront Park Expansion." *NJ.com*. October 05, 2016.
The Newark Experience: Shaping the City, Rutgers University Libraries, Library Guides. retrieved from http://libguides.rutgers.edu/.

Part II

Artificial Spaces

5 Passages

From Artificial Animal to Planetary Man

Paolo Perulli

This chapter reflects upon the modes through which Western socio-philosophical discourse has critically explored Nature and the City. The essay examines the writings of a number of authors, ranging from Walter Benjamin to Bruno Latour, i.e. European critical thinkers of modernity from the 1920s' critical turn to present debates. Two crises of modern capitalism, the 1920–1930s and the 2007–ongoing crises, are the alpha and omega of this critical discourse. During this period, globalization processes have increased dramatically and a planetary urbanization of the world has emerged. In the course of the 20th century much has changed in the relation between Nature and the City, as their modes of representation dramatically shifted. Nature has been fully subsumed and incorporated into the Artificial, whose main representation is the City in its metropolitan and, today, postmetropolitan forms.

More precisely, the aim of this chapter is to show how images of the City as 'artifact,' which surfaced in the 1960s–1970s—in parallel to the postmodern turn and the rapid planetary urbanization of the world, have been progressively detached from their original reference to Nature. This detachment is fully expressed in texts such as French sociologist Guy Debord' s *Society of Spectacle* (1967) and French philosopher Jean-Francois Lyotard's *The Postmodern Condition* (1979). The origins of this representational detachment can be traced (back) to German philosopher Walter Benjamin's work on *Passages* (*Passagenwerk* in German) and therefore are profoundly rooted in a critical reflection on early modernity.[1] *Passages* is Benjamin's most important intellectual output written during his exile in Paris in the 1930s—before his death at the French-Spanish border in 1940. Never completed, the *Passagenwerk* was rediscovered and translated later and has widely influenced (from Adorno to Arendt) the philosophical and aesthetic discourses in the second half of 20th century. I will also suggest in the conclusion to this reflection that the City's detachment from Nature is likely to be predated to the very origin of Western political modernity, i.e. to Thomas Hobbes' *Leviathan* (1651), the foundational treatise in the Western world on State sovereignty.[2] Since then the State has subsumed the force of Nature (as well as that of Religion, and of any other power on the Earth) into its 'artificial' power.

If Walter Benjamin's Parisian *Passages* were the 19th-century privileged architectural model of the modern urbanite, consumer and *flaneur*, the Planetary Man of contemporary globalization is still waiting for a new *Passage* to take him towards planetary urbanization. No single architecture can fully express or articulate such a planetary condition: 'postmetropolis' remains for now the most convincing, if not precise definition of the new Urban Epoch. It is an Epoch of loneliness, of a structurally 'liquid' social modernity highlighting at the same time the fast changes occurring on uneven planetary scales. Walter Benjamin (1921) wrote that: [3]

> This passage of 'planetary man' (Planeten Mensch) through the house of despair is, in the absolute loneliness of his path, the ethos Nietzsche describes. This man is the Ubermensch, the first who knowingly begins to realize the capitalist religion.[4]

Planetary Man is the protagonist of our contemporary 'liquid' society. The German philosopher Ernst Junger later added (in 1951) that only '*Der Waldgang*' remained to the free individual in contemporary society. The 'retreat in the forest,' anti-urban as it is, is the only exit choice we have in the Epoch of planetary anxiety and fear. For the image of the wounded Planet, which has been multiplied by ecological thought since the 1960s/1970s, is now deeply embedded in our contemporary view of the future.

Modernity

In the archives forming the background of his unfinished book, *Paris—Capital of the 19th Century*, Walter Benjamin collected material on "Nature and the City," a project documented by Giorgio Agamben in his monumental study on Benjamin (2012).[5] The chthonic nature of the city, its underground base and its submarine image are among the favored themes to appear in the fragments collected by the German thinker of modernity. In Benjamin, the ruins are signs of the dialectic between the technological growth of the metropolis and the destruction made possible by technological advance itself. This was a departure from the view of Georg Simmel, one of Benjamin's masters whose images of the ruins are a product of man destined to become a product of Nature, a cosmic tragedy. Cities (from London to Lisbon to Tokyo in the past centuries) are always in danger of destruction by water, fire and other natural catastrophic events as well as by social and technological changes, exemplified by the urban transmutation of Hausmannian Paris. "The city of Paris has entered the 20th century in the form that Haussmann has given to it," Benjamin wrote. Haussmann achieved subversion of the Paris' urban image through such simple tools as spades, hoes and picks. "What destruction—Benjamin observed—such limited tools have produced! And how since then, in the epoch of great cities, the instruments for razing them have improved! How they evoke images of the future!"[6] The urban future is, according to Benjamin, made of ruins (*Trümmern* in German) and destruction, just as those that the tempest of Progress constantly produces (discussed in the last writings of Benjamin, the *Theses on the Philosophy of History*). Planetary urbanization in that respect can be understood as simply transposing such destruction onto a global scale.

Our starting point is therefore Paris in 1859, the first year of Haussmann's public works, which can be considered as emblematic of the joint processes of urban change, urban renewal and urban destruction. Imitating Haussmann on an ever larger scale was to become the dream of every urban *growth machine*; criticism of Haussmann's vision would become the leitmotiv of any critical approach to urban modernity, including that of the planetary urbanization theorized by critical theorists such as the French philosopher Henry Lefebvre and the American sociologist Neil Brenner.[7]

In the very same years (1848–1870) that saw Haussmann at work, we can say that the entire history of modernity can be detected. The European revolution-counterrevolution dialectic is incised deep in the streets of Paris. What comes first, and what follows this crucial Epoch is encapsulated in the image of *Paris, Capital of the 19th Century*. Benjamin was fully aware of this, and meticulously collected evidence of Paris' urban change, its preparation and its consequences. The apparatus for destruction is central in Charles Baudelaire's poetic work, which Benjamin saw as the lyrical acme in the epoch of mature capitalism: "the devaluation of the human world caused by the market economy deeply interacts in its historical experience."[8] And also "Baudelaire's lyrics are located at the point where the nature of things is overwhelmed and transfigured by the nature of man."[9]

Along the axis designating the structural-technical form of the city, we can "locate" the following questions in Benjamin's inquiry: in which period does the expression "a sea of houses" make its first appearance? when did Haussmann's demolitions begin? And the following points: a reflection on the traffic of carriages on the Pont Neuf; on the appearance of the first mailboxes in Paris; on the arrival of the electric telegraph, on the development of administrative control; and the advent of house numbering.

Since the mid-19th century, the modern metropolis has become the "field" where all sorts of experiments in urban population control are conducted by means of technical instruments and flow regulation. Here the art of government made its first appearance: during the 18th century in the pre-revolutionary *ancien regime* institutions, in the 19th century in the new forms of urban modernity. But Paris is also the city whose streets are occupied by practices in which tradition and urbanity interpenetrate. Along the axis which designates the subjectivity of the urban spirit, we can therefore "locate" the following points in Benjamin's inquiry: the *inhabited interior*; the *passages*; the *crowd, mass, flâneurs*; the *attitude (of intellectuals, bohemians)*; an attention to *psychoanalysis, and the phantasmagory of merchandise*.

In a text by Baudelaire discussing the work of Charles Meyron, a painter and a contemporary of his in Paris, the interconnectedness of antiquity and modernity is clearly underlined. "Rarely have we seen represented with such poetical richness the natural solemnity of a capital city" writes Baudelaire in *Ecrits sur l'art*: stratified stones and industrial obelisks, monumental restoration works are mixed together in cobwebs fashion. Antiquity and modernity are interconnected in the new city. This vision of modernity is the dialectical basis of the *Passages*, which are considered by Benjamin to be both architecture and symbol, representing the dream of the urban collectivity.

The protagonist here is a particular kind of urbanite, the *flaneur* in the endless flows of urban population. Edgar Allan Poe's "*The Man of the Crowd*" is the explicit model:[10] but in Poe no *flanerie* is admitted; rather, the protagonist is forever within the flow. Yet Benjamin ironically notes: "The habitus of the flaneur who botanizes on the asphalt."[11] Again, the dialectic here is between man, nature and commodity. Nature is never seen as a value in Baudelaire: it is seen as something saddening, hard and cruel.[12] The cruelty of nature may be the reason why he refused communion with any terrestrial nature. It is the city which attracts the poet: its labyrinth, its mosaic of ruins, its sky.[13] For Benjamin, however, the Parisian *Passages* are also the artificial spaces where the emerging market economy and the display of its fashions and commodities surface. Places which—as later will be true of the giant shopping malls, airports, *non-lieux*—we pass through without any personal, historical or identitarian commitment or promise. In this sense the 19th-century Parisian *Passages* are the prototype of the urban destruction of any natural environment caused by the inherent capitalist logic, which was to be later thematized by ecological thought in the following century.

Postmodernity

In postmodernity, the roles of Nature and the City dramatically change as the notion of 'paysage,' the landscape ceases to exists. Landscape is now, as articulated in British geographer David Harvey's critical account of *The Condition of Postmodernity* or in American novelist Don De Lillo's *Cosmopolis*, a built-up environment.[14] The dialectic turns into an opposition between Economy (*oikos nomos*) and Ecology (*oikos logos*). Both economy and ecology derive their meaning from *oikos*, the domestic symbol of human permanence on the Earth. The strength of the symbol is again documented in the influential Swiss psychoanalyst Carl Gustav Jung's *Red Book* written between 1914 and 1930:[15] the house as a symbol of stability in the Mandala circle, one of the oldest religious symbols analyzed by Jung (significantly, it seems, in 1937 Benjamin was thinking of Jung and Klages as his possible next research work, unfortunately never carried out).[16] Ludwig Klages' text entitled "Man and the Earth" (*Mensch und Erde*, 1913)[17] is clearly ahead of its time: it anticipates our contemporary ecological view and is now considered a preview of ecological thinking. For according to Klages, Humanity is destroying Nature through its progress. Furthermore, as critical thinking on the planetary urbanization of the world suggests, the contemporary neoliberal city is the ideal place for such annihilation of Nature. The capitalist imperative is dominant today: the city finances the city through the urban rent in permanent, endless urban expansion. Masses, or multitudes are no longer the protagonists here. Only the artificial man produced through technological progress can be the protagonist. Benjamin called him Planetary Man (Planeten Mensch). And this man is a believer in a new religion, capitalism.

At this point, the inside–outside dialectic, which is typical of modernity comes to an end: in the passage from modern to postmodern and from imperialism to Empire there is progressively less distinction between inside and outside (according to Hardt and Negri).[18] This is particularly true as far as the dialectic of sovereignty between the civil order and the natural order comes to an end. As American critic Fredric Jameson puts it:

> In modernism (. . .) some residual zones of "nature" or "being", of the old, the older, the archaic, still subsist; culture can still do something to that nature and work at transforming the "referent". Postmodernism is what you have when the modernization process is complete and nature is gone for good.[19]

There is no 'outside' here any longer. In the transition from modern to postmodern thinking there is progressively less distinction between inside and outside due to capitalism's full appropriation of the world space. Under planetary urbanization, urban theory remains without an outside—as Neil Brenner points out: the complete transformation of the Globe into an urban realm has been virtually concluded.[20] In the coming era of virtuality the City becomes the realm of the Artificial, both in terms of physical planning and in terms of architecture. In urban planning, technological tools are employed to produce simulations of the city. First of all comes the map and its rendering techniques: as French philosopher Jean Baudrillard puts it,[21] reality is the product of miniaturized units, of matrices, databases and models of command—in this way reality can be reproduced an infinite number of times. Importantly we should always distinguish between *Realität* and *Wirklichkheit*: the former term defines the fiction we build to find stability in the world, the latter defines the chaotic flux of the world, without beginning and without end.

In the field of urban architecture, the 'artificial' takes the form of the 'foam-city' described by German philosopher Peter Sloterdijk in *Shaume, Spheren III*.[22] Again, a natural, biological concept, foam, is employed here to define the contemporary human condition: the cellular housing unit gives rise to repeated massive foam, in which 'free' individuals, flexibly adapting to the flow of capital accumulation, develop their intimacy. But it is a lost intimacy, as the liberation of individuals living alone with the aid of individualizing techniques of habitat and mass-media is fully based on market relations. The modes of representation of human environments are still based on Nature, but only as manipulated artificial products of a post-urban or even post-human Epoch fully dominated by technology and market logics.

Bridges

This vision is close to that of the Nietzschean 'last men' presented in the Prologue of *Also Spracht Zarathustra*:

> We have invented happiness, say the last men and they blink. They have left the regions where it was hard to live, for one needs warmth. One still loves one's neighbor and rubs against him . . . One still works, for work is a form of entertainment. One no longer becomes poor or rich: both require too much exertion. Who still wants to rule? Who obey? Both require too much exertion.[23]

To overcome this final meta-historical outcome we need a renewed dialectical approach, which can be proposed only by considering new models of circularity bridging human actors and Nature. This bridge is only possible if we follow philosophical-anthropological lessons such as the one proposed by French philosopher Simone Weil's supernatural knowledge ('*connaissance surnaturelle*'): the city is not 'the social,' it is the air that we breathe, our roots and traditions.[24] In cities like Troy or Venice (two of Weil' s models) such supernatural knowledge is clearly expressed. In a tragedy Simone Weil wrote in the last years of her life, *Venise sauvée* (Venice saved) the heroic protagonist who saves Venice from the foreign conspiracy is a noble witness of such spirit of the city (nature, past, tradition).[25] The same spirit still remains impressed in the inhabitants leaving their city of Troy (defined 'sacred'): migrants like Aeneas are the prototype of the European tradition of 'rootedness' (hence we Europeans closing our borders to foreigners today have forgotten our very same origin).

In the field of the built-up environment, we can follow in the same vein German architect Bruno Taut's glass-architecture as a symbol of supernatural knowledge: it is the idea of a town which will survive modernization and urban sprawl, the dispersed community of "*The Dissolution of Cities*" (1920).[26] Taut bridges the divide between Nature and the City through an environment that couples the natural landscape of the Swiss Alps with man-made building technology. It does not matter that Taut designed glass utopian domes to be superimposed on the Alps: his theoretical contribution goes beyond architecture itself. In his Utopia for the Third Millennium, as *The Dissolution of Cities* seems to us, work cooperatives, agricultural and dispersed production centers will arise. No centralization is admitted; rather, massive decentralization is advocated. A region of 'glass' gardeners, ecological homes with water reservoirs and mobile solar radiators. The more distant men are from each other in space, the more they develop hospitality. A community of individualities develops circular forms of society, in which crowns, flowers, stars are forms of Taut's imagined community—a transpassing, transparent society, embodied by the glass-architecture. It is an astonishing foretaste of our possible coming future: the end of gated territories and borders, free spaces against closed spaces. We should subvert our architecture to enhance the level of our civilization, as Paul Sheerbart (one of Taut's masters) asserts in his *Glas Architektur* (1914):[27] "and this will be possible only by eliminating the closures of the spaces in which we live." We are here beyond the old *Nomos of the Earth* (the text of international law written by German philosopher Carl Schmitt in 1950):[28] the ring, the fence as forms of the political and juridical community will be surpassed in the framework of people's peaceful coexistence on the Planet.

Planetary Urbanization

One century later, the openness of spaces and new urban constellations have taken a different direction than that foreseen by Taut and Sheerbart. Gated communities are growing in the United States, and in China as well.[29] Global migration from the country to the city and from poor to rich countries across the Planet is increasing. The geography of globalization and planetary urbanization have radically changed the world. However, the globalization process is more controversial than expected. It has also been interpreted as the final outcome of a 'lost geography,' formed during the last century of the 'American Empire.'[30]

> To the extent that the geography of the American Century remains obscure, the origins, outlines, possibilities, and limits of what today is called globalization will also remain obscure. There is no way to understand where the global shifts of the last twenty years came from or where they will lead without understanding how, throughout the twentieth century, U.S. corporate, political, and military power mapped an emerging empire.[31]

Ideas of a spaceless and unlimited 'Empire' have also been elaborated as a passage from imperialism to a new universal order that accepts no boundaries or limits, driven by a ubiquitous, multi-layered capitalist system towards a global multitude of individuals across the world.[32] However the globalization process seems to be more dispersed and multipolar than both Smith's 'American Empire' and Hardt and Negri's 'Empire' are able to represent. The urban realm is of the utmost importance if we are to understand this new acentric morphology. Geographers like Edward Soja and Miguel Kanai presented our urban world in these terms:

> Over the past 30 years, the world has been experiencing an unusually expansive and reconfigured form of urbanization that has been defined as a distinctively global urban age—one in which we can speak of both the urbanization of the entire globe and the globalization of urbanism as a way of life (. . .). We see this extended form of contemporary urbanization not just as an adjunct to the globalization process, but also as its primary driving force, stimulating innovation, creativity and economic growth while at the same time intensifying social and

economic inequalities and conflict filled political polarization. But as the world urbanizes, cities are being globalized.[33]

How can we react to such extensive, complete and uneven urbanization of the world? In terms evoking Walter Benjamin's theses, Neil Brenner prompts us to realize that:

> a new understanding of planetary urbanization may prove useful to ongoing struggles against neo-Haussmannization, planetary enclosure, market fundamentalism and global ecological plunder.[34]

It is interesting to find in urban critical theorists such Haussmann-like descriptions of the current urbanization of the Earth. The Globe and the Earth are now opposed: on one side we find the last product of human creation (the Globe), on the other side the last version of Nature (the Earth). However, the City as opposed to Nature is now detached from its original urban form and assigned to a generalized formation of urbanization. In the collective volume edited by Brenner under the Lefebvrian title of *Implosions/ Explosions*, a text by the Swiss sociologist Christian Schmid is devoted to the disintegration of the hinterland and the end of wilderness in the Swiss context. On this case it is also worth taking a look at Roger Diener, Jacques Herzog, Marcel Meili, Pierre de Meuron and Christian Schmid's *Switzerland-An Urban Portrait*,[35] where the complete urbanization of the Swiss territory is contrasted to the traditional view of Switzerland as a realm of Nature. Metropolitan regions and networks of cities dominate: according to Schmid "50 years ago the countryside still represented a continuous backdrop against which the urban centers stood out; now, by contrast, the last remains of rural areas are like a patchwork of solitary islands" and "Alpine resorts are urban regions in the mountains."[36]

A possible way to react to such extensive urbanization of the Planet is proposed by the architects, urban planners and progressive urban coalitions engaged in "doing nature" as a product of urban land reuse. The "discursive way of doing nature," as in the case of River Elbe in Hamburg or other case studies on green cities and eco-cities in Germany, Denmark and elsewhere, is presented as a new approach which allows for constructing nature as a result of "contested natures."[37] Authors adopting this perspective suggest identifying the social and cultural contexts within which environmental concepts are formed. Nature here is 'invented' by environmentalism, it is 'produced' as countryside, it is 'reconfigured' within contemporary policy and politics, and it is governed as in the case of great environmental risks.[38]

This is a paradoxical reversal of Haussmann's perspective: instead of destruction, urban renewal here is advocated as the re-creation of nature. Many fragmentary initiatives, from the ecological reuse of already used urban land to urban reconstruction of agricultural spaces, go in the very same direction. It is hard to say if this is a possible reversal of more robust trends towards the unlimited urbanization of the Planet foreseen and theorized by Lefebvre in the 1970s, or if it is possibly a limited, 'niche' answer to the currently diffused social demands of *governing the commons* such as land, water, nature in a less disruptive way.[39]

The British geographer Peter Taylor's argument runs in the same direction on a vaster scale. In the conclusion of his latest book, *Extraordinary Cities*,[40] he demonstrates that endogenous and exogenous limits to our world system are calling for a new answer to the 21st century as a crisis century. A renewed version of the 'garden city of tomorrow' and the New Town movements of the 20th century are advocated. According to Taylor (influenced by Jane Jacobs' legacy), modern green networks of cities are among the possible answers. Modularity and alternative import replacement are part of the recipe. Green city networks would be the next urban frontier of the 21st century, Taylor concludes optimistically:

> We do not know what this is to be but we can say that guardians will be indispensable to thinking about reaching a sustainable future. Guardians may work through multiple new states, or through just one global state, or through a new form of global governance not yet forseeable.[41]

In his recent research work on modernity, on the other hand, the German critical thinker Peter Wagner has taken a more skeptical approach on political-philosophical grounds.[42] Well aware of Benjamin's critical vision of historical progress and elaborating on Claus Offe's critical studies, Wagner addresses the question as to which concept of political progress is still sustainable today. His conclusion is skeptical. Given the disappointment with past progressive efforts and the highly problematic current global conditions—threats of war, energy scarcity and climate deterioration—the most we can realistically aim at is avoiding regression, rather than any progressive political improvement or enhancement. It is the same argument as that developed by Carl Schmitt in his masterpiece, *The Nomos of the Earth* (1950).[43] Past so-called 'conservative' thinkers as well as contemporary 'radical' thinkers seem to converge in the same perspective. Their pessimistic conclusion runs:

> The new global space (no longer "spatial" at all) fundamentally has many heads and is marked by an endemic conflict among impersonal powers that no longer express any value.[44]

The Italian philosopher Massimo Cacciari (2014) has also returned to Leon Battista Alberti's Renaissance symbols. In his preface to *The Winged Eye. Migration of a Symbol*[45] he underlines that Alberti's winged eye can see all human vicissitudes, the constant tension between fortune and virtue, the irremediable vices of human nature. But ultimately the winged eye has to recognize that *physis* (Nature) likes to hide, and the Intellect's path is too deep to be fully represented. *Physis* (Nature) is in fact an uncontrollable, incalculable mechanism.[46] The etymology of *Physis*, from the ancient Greek verb *phyo*, means 'to generate," to grow.' The wholly artificial production of today's human knowledge is made of a different substance than Nature. *Homo faber* has finally canceled Nature. But *Physis* will continue to represent an alternative mode of representation of the world. When human civilization will come to an end, Nature will continue to generate, to grow. The artifact, considered as any object produced by human craft, be it a tool, weapon or ornament, will be reclaimed by Nature.

The New Architecture of Passages

Contemporary architecture still has to find its own 'City of Passages,' yet some leading architects such as Rem Koolhaas have suggested airports and other mobility infrastructures could be our cities of passages. We are simply inhabiting spaces of transition. The city of flows designed by Koolhaas is at the moment only an analytical tool in the thinking of contemporary sociologists such as Manuel Castells and Bruno Latour.[47] In his influential urban theory, the former has defined the "space of flows" as an artifact able to progressively substitute itself to the old "space of places." The latter in particular has developed a theory of Actor/Network (ANT), as well as a theory of Inquiry into Modes of Existence (AIME)[48] which challenges modernity and calls for a new anthropology. '*Reset modernity!*' is Latour's latest manifesto. Latour (2016) criticizes the assimilation of Globe and Earth as a product of European thinking based on the notion of Empire in political studies. Underlying this vision of globalization is the reduction of entities, including Nature and natural objects, to localization. Localizing in a map means that "we accept to begin from localization in time and space, this land, this site, this moment, this city, this individual agent, but we never trace our way back to the formidable instrument that enables the attribution of coordinates." Localization is a form of representation, in the double significance of 'localizing' and 'ordering.' This instrument, able to offer localization to any chunk of space-time, is again qualified as a product of European thinking. It is connected to the concept of sovereignty, which can be traced back to the Roman Empire and to a principle of localization seen as the definition of boundaries. Relation, scale, causality and agency are the keywords of the Latourian critical theory of sovereignty. Relations of an object, subject and so on with other entities (individuals, States, economies) are defined by the principle of localization only once object, subject and so on have been defined as individual entities. Of course this means that entities are given their characteristics prior to considering their 'relations,' when in fact these relations are essential to entities' qualities. This process is an 'artifact,' a phantom that no amount of relation can resuscitate afterwards. *Homo economicus* is the outstanding case of the artificial creation of a subject which occurs without taking into account the set of relations within which he is surrounded and the influences of the environment in which he is embedded. This is true in social theory, but also in biology or geopolitics, etc.

The other 'artifact' created by the principle of localization is the idea of scale (geographical, political), understood as the localizing of an object or subject along a continuum from local to global. This is somewhat artificial, as if, looking at the 'small' (e.g. a certain site), we were driven by a logic of localization, prompted to say that the local site is to be situated in a 'large' scale. The principle of localization refers here to a representation of scale as 'given,' but on the contrary we need to reorient our analysis to the processes of scaling and rescaling.

The third 'artifact' created by the localization principle is causality, which is the attempt to connect atomized, isolated entities through cause-effect chains transforming engineering principles of reality into metaphysics. And, finally, we have the 'artifact' of agency. The isolated entity cannot have any agency unless connected to other parts, but in accordance with the general laws of nature. Here the physical sciences have—according to Latour—adopted a model designed by the political sciences of sovereignty. It is the Hobbesian theory of the State, in which a single entity without any autonomous agency (the individual as parcel of an amorphic multitude) is subjected to a larger entity, the State.

It is surely significant that Thomas Hobbes calls the State an artificial animal (*artificiale animal*). The beginning of "Leviathan" is worth quoting:[49]

> Nature (the art whereby God hath made and governs the world) is by the art of man, as in many other things, so in this also imitated, that it can make an artificial animal. For seeing life is but a motion of limbs, the beginning whereof is in some

> principal part within; why may we not say, that all automata (engines that move themselves by springs and wheels as doth a watch) have an artificial life? (. . .) For by art is created that great LEVIATHAN called a COMMONWEALTH, or STATE (in Latin CIVITAS) which is but an artificial man.[50]

The 'artifact' created by the art of man imitating Nature, State sovereignty, is ultimately destined to destroy Nature. Artificial reason and will, which are according to Hobbes a product of human creation similar to God's creation, are destined to turn the Globe into an inhabitable dreamland of metaphysics, as Latour puts it. Or into the house of despair, as Benjamin would have concluded.

Humanity has produced 'artifacts' (etymologically: the making of arts) as valued objects of art destined to last in time, according to the Greek philosophy of art as techne (τέχνη). But today's technological progress, the myth of our Epoch, is our necessity and at the same time our destiny. Technology has taken the lead but the direction taken by technological progress is obscure. Any other power, religious or political or ethical, is today subsumed and incorporated into the technical power. The most important legacy of the Western culture, Plato's dialogs, assert that the Sky and the Earth, the Gods and the humans are united by common (*koinòn*), friendship, temperance and justice. This is why, Plato asserts in *Gorgias* (508 A), we call the entire universe a *kosmos* (order). We have simply forgotten, in the Epoch of Planetary Man, these 'cosmic' modes of representation. Getting global we have lost the order. But our world needs an order to avoid chaos. The increasing urban extension and global technical interconnectedness of today's world are simply epiphenomena of the forms assumed by such mode of representation that we call globalization. The new architecture of passages needs not only to facilitate and speed the transit of people moving around the world, but also to erect new bridges, political and cultural, across the diverse populations of the world.

Conclusions

The very nature of globalization, its linking of the diverse populations of the Globe, cannot be reached unless we introduce other categories of thinking in tension with current technological progress: namely, lawfulness and institutional solidarity. Law, solidarity and technological progress are the polar stars of a world system to be designed and created anew. In order to do so, we can recapitulate the trajectory we have followed since here. The passages, the Arcades project of Benjamin has been presented in the first part of the chapter and then projected from modernity to postmodernity. In this projection Nature and the City turn to an opposition between Ecology and Economy, both derived by the Greek word *oikos*. This term was the symbol of our stability on the earth. In Greek culture, two Gods, Hestia (the domestic Goddess) and Hermes (the God of communication and mobility), were the custodians and guardians of *oikos*: fixity and motion were interconnected and allied. Today Earth, Globe and Land are detached and opposed. Earth is the domain of nature and ecology; Globe is the political vision of sovereignty under globalization; Land is the amorphous ground for economic exploitation and planetary urbanization. The Latourian Modes of Existence (including law, politics, religion, fiction, attachment, organization) are the epistemological ground for a new alliance and show that the idea of globalization (i.e. Western modernization spread over the whole Planet) has come up against unexpected opposition from the Planet itself. This 'alternative' mode of representing globalization is of the utmost importance today. It means that Nature (*Physis*) and the City (the human cohabitation on the Earth) need new common ground to be found: a new search, or discovery, different than the modern project which now—in times of global disorder and clash—comes to an end. A new approach to 'cosmic' modes of representation able to reunite again the double meaning of *kosmos*, globe and order, would be needed.

Notes

1 Walter Benjamin, *Das Passagenwerk*, (Frankfurt am Main: Suhrkamp Verlag, 1982).
2 Thomas Hobbes, *Leviatano* (or. ed.: *Leviathan*, 1651) (Milan: Bompiani, 2001).
3 Walter Benjamin, *Capitalism as Religion*, [Fragment 74]1, Translated in English by Chad Kautzer from the *Gesammelte Schriften* (1985), (Frankfurt: Suhrkamp Verlag, 1921).
4 Benjamin, *Capitalism as Religion*, 260.
5 Walter Benjamin, *Charles Baudelaire. Un poeta lirico nell'età del capitalismo avanzato*, (or. title: *Das Paris des Second Empire bei Baudelaire*), in Giorio Agamben, Barbara Chitussi and C.-C. Haerle (eds.), *Charles Baudelaire* (Vicenza: Neri Pozza, 2012).
6 Benjamin, *Charles Baudelaire. Un poeta lirico nell'età del capitalismo avanzato*, 700.
7 Henri Lefebvre, *La révolution urbaine* (Paris: Gallimard, 1970); Neil Brenner (ed.), *Implosions/Explosions* (Berlin: Jovis Verlag, 2014).
8 Benjamin, *Charles Baudelaire. Un poeta lirico nell'età del capitalismo avanzato*, 58 (author's translation).
9 Benjamin, *Charles Baudelaire. Un poeta lirico nell'età del capitalismo avanzato*, 59 (author's translation).
10 Edga Alan Poe, *The Man of the Crowd*, in *Tales and Sketches: To Which Is Added the Raven: A Poem* (London: George, Routledge & Co, 1852), Poe's text is quoted in Benjamin, *Charles Baudelaire. Un poeta lirico nell'età del capitalismo avanzato*, 666.
11 Benjamin, *Charles Baudelaire. Un poeta lirico nell'età del capitalismo avanzato*, 655 (author's translation).
12 "J'ai même toujours pensé qu'il y avait dans la nature florissante et rajeunie, quelque chose d'affligeant, de dur, de cruel-un je ne sais quoi qui frise l'impudence » (quoted in Walter Benjamin, *Charles Baudelaire. Un poeta lirico nell'età del capitalismo avanzato*. 153).
13 "Le grands ciels qui font rêver d'éternité" Baudelaire writes in a poem entitled *Paysage* (*Le Fleurs du Mal*, LXXXVI).
14 David Harvey, *The Condition of Postmodernity* (Oxford: Blackwell, 1991); Don De Lillo, *Cosmopolis* (New York: Scribner, 2003).
15 Carl Gustav Jung, *The Red Book. Liber Novus* (New York: W.W. Norton & Co., 2010).
16 Richard Wolin's *Labyrinths: Explorations in the Critical History of Ideas* provides the best summary of Benjamin's proposed essay on Jung and Klages. Benjamin's concept of the "collective unconscious"—one of the methodological keys to the Arcades Project—was explicitly derived from Jung. Benjamin later came to view a theoretical self-clarification vis-a-vis Jung as an imperative task. He alludes to this project in a 1937 letter to Scholem: "I wish to secure certain methodological fundaments of 'Paris Arcades' via a confrontation with the theories of Jung—especially those of the archaic image and the collective unconscious." His proposed study of the differences separating his utilization of these concepts from that of Jung and Klages was rebuffed by the Institute for Social Research. This is possibly due to the refusal of mitical thinking of Klages and Jung by Adorno and others marxist intellectuals of the Institute. [Wolin, *Labyrinth*. 70. http://books.google.com/books?id=FdkQrXnPpUQC]
17 Ludwig Klages, *Mensh und Erde* (Berlin: Matthes § Seitz, 2013).
18 Michael Hardt and Antonio Negri, *Empire* (Cambridge-London: Harvard University Press, 2000), 187.
19 Fredric Jameson, *Postmodernism, or, the Cultural Logic of Late Capitalism* (Durham: Duke University Press, 1991).
20 Brenner, *Implosions/Explosions: Towards a Study of Planetary Urbanization*.

21 Jean Baudrillard, *The Consumer Society: Myths and Structures* (London: Sage, 1998).
22 Peter Sloterdijk, *Shaume, Spheren III,* (Frankfurt: Shurkamp Verlag, 2005).
23 Friedrich Nietzsche, *Thus Spoke Zarathustra*, Walter Kaufman translation (or. 1883–5), 3–5 www.theperspectivesofnietzsche.com/nietzsche/nuber.html
24 Simone Weil, *Cahiers,II* (Paris: Librairie Plon, 1972).
25 The point is clearly expressed in the pages of Cahier VII, and also in other unfinished texts like *Venise sauvée.*
26 Bruno Taut, *La dissoluzione della città* (Faenza: Faenza Editrice, 1976) (or. title: *The Dissolution of Cities*, 1920).
27 Paul Sheerbart, *Architettura di vetro*, (Milano: Adelphi, 1982), (or. title: *Glas-Architektur*, 1914).
28 Carl Schmitt, *Der Nomos der Erde* (Berlin: Duncker § Humblot, 1974, or. 1950).
29 An in-depth analysis of recent creation of urban closed communities in China is in Youqin Huang and Li Si-ming eds., *Housing Inequality in Chinese Cities* (London-New York: Routledge, 2014); Wang Xiaoming, *Città senza limiti. Studi culturali sull'urbanizzazione cinese* (Venezia: Libreria Editrice Cafoscarina, 2016).
30 Neil Smith, *American Empire* (Los Angeles-London: University of California Press, 2003).
31 Smith, *American Empire*, 4.
32 Michael Hardt and Antonio Negri, *Empire.*
33 Edward W. Soja and Miguel Kanai, "The Urbanization of the World", in Neil Brenner (ed.), *Implosions/Explosions: Towards a Study of Planetary Urbanization*, 142.
34 Neil Brenner, "Urban Theory Without an Outside", in Neil Brenner (ed.), *Implosions/Explosions: Towards a Study of Planetary Urbanization*, 28.
35 Roger Diener, Jacques Herzog, Marcel Meili., Pierre de Meuron, and Christian Schmid, *Switzerland-An Urban Portrait* (Basel: Birkhauser, 2006).
36 Christian Schmid, "The Urbanization of Switzerland," in N. Brenner (ed.), *Implosions/Explosions*, 268–275.
37 Phil Macnaghten and John Urry, *Contested Natures* (London: Sage, 1998).
38 Sybille Bauriedl and Sabine Hohler, *Doing Nature: River Elbe under Discursive Construction*, Paper presented at the 2003 Hamburg conference: "Does Discourse Matters: Power, Discourse and Institutions in the Sustainability Transition", Germany, Hamburg, Hauss Rissen, 11–13 July 2003. See also the recent International Conference on Green Urbanism, Rome (October 2016) to be published by Elsevier.
39 Elinor Ostrom, *Governing the Commons* (Cambridge, UK, New York: Cambridge University Press, 1990).
40 Peter J. Taylor, *Extraordinary Cities* (Cheltenham: Edward Elgar, 2013).
41 Taylor, *Extraordinary Cities*, 378.
42 Peter Wagner, *Modernity. Understanding the Present* (Cambridge: Polity Press, 2012).
43 See the already quoted text of Schmitt.
44 Massimo Cacciari, *Europe and Empire: On the Political Forms of Globalization* (New York: Fordham University Press, 2016), 156.
45 Massimo Cacciari, "Preface" to Alberto Giorgio Cassani, *L'Occhio Alato. Migrazioni di un simbolo* (Torino: Aragno, 2014).
46 Jacques Derrida, *Athens, Still Remains* (New York: Fordham University Press, 2010).
47 Bruno Latour, "Onus Orbis Terrarum. About a Possible Shift in the Definition of Sovereignty," in *Millennium: Journal of International Studies*, 44 (2016), Special issue on Failure and denial in World Politics.
48 Bruno Latour, *An Inquiry Into Modes of Existence* (Cambridge, MA: Harvard University Press, 2013).
49 Hobbes, *Leviatano.*
50 Hobbes, *Leviatano*, 1.

References

Baudrillard, Jean. *The Consumer Society: Myths and Structures*. London: Sage, 1998.
Bauriedl, Sybille, and Sabine Hohler. *Doing Nature: River Elbe Under Discursive Construction*. Paper presented at the 2003 Hamburg conference: "Does Discourse Matters: Power,

Discourse and Institutions in the Sustainability Transition", Germany, Hamburg, Hauss Rissen, 11–13 July 2003.
Benjamin, Walter. *Capitalism as Religion*, (1921) [Fragment 74]1, Translated in English by Chad Kautzer from the *Gesammelte Schriften*. Frankfurt: Suhrkamp Verlag, 1985.
Benjamin, Walter. *Das Passagenwerk*. Frankfurt am Main: Suhrkamp Verlag, 1982.
Benjamin, Walter. "Charles Baudelaire. Un poeta lirico nell'età del capitalismo avanzato, (or. title: Das Paris des Second Empire bei Baudelaire)." in Giorgio Agamben, Barbara Chitussi and C.-C. Haerle (eds.), *Charles Baudelaire*. Vicenza: Neri Pozza, 2012.
Brenner, Neil (ed.). *Implosions/Explosions*. Berlin: Jovis Verlag, 2014.
Cacciari, Massimo. *Europe and Empire. On the Political Forms of Globalization*. New York: Fordham University Press, 2016.
Debord, Guy. *La Societé du Spectacle*. Paris: Buchet-Chastel, 1967.
Lillo, Don De. *Cosmopolis*. New York: Scribner, 2003.
Derrida, Jacques. *Athens, Still Remains*. New York: Fordham University Press, 2010.
Diener, Roger, Jacques Herzog, Marcel Meili, Jean de Meuron, and Christian Schmid, C. *Switzerland-An Urban Portrait*. Basel: Birkhauser, 2006.
Hardt, Michael, and Antonio Negri. *Empire*. Cambridge-London: Harvard University Press, 2000.
Harvey, David. *The Condition of Postmodernity*. Oxford: Blackwell, 1989.
Hobbes, Thomas. *Leviatano* (or. ed.: *Leviathan*, 1651). Milan: Bompiani, 2001.
Jameson, Fredric. *Postmodernism, or, the Cultural Logic of Late Capitalism*. Durham: Duke University Press, 1991.
Jung, Carl Gustav. *The Red Book. Liber Novus*. New York: W.W. Norton & Co., 2010.
Klages, Ludwig. *Mensh und Erde*. Berlin: Matthes § Seitz, 2013.
Latour, Bruno. *An Inquiry Into Modes of Existence*. Cambridge, MA: Harvard University Press, 2013.
Latour, Bruno. "Onus Orbis Terrarum. About a Possible Shift in the Definition of Sovereignty." *Millennium: Journal of International Studies* 44, Special Issue on Failure and Denial in World Politics, 2016.
Lefebvre, Henri. *La révolution urbaine*. Paris: Gallimard, 1970.
Lyotard, Jean-François. *La Condition Postmoderne*. Paris: Les Editions de Minuit, 1979.
Macnaghten, Phil, and John Urry. *Contested Natures*. London: Sage, 1998.
Nietzsche, Friedrich. *Thus Spoke Zarathustra*. Walter Kaufman translation, (or. 1883–5). retrieved from www.theperspectivesofnietzsche.com/nietzsche/nuber.html.
Ostrom, Elinor. *Governing the Commons*. Cambridge, New York: Cambridge University Press, 1990.
Poe, Edgar Allan. "The Man of the Crowd." in *Tales and Sketches: To Which Is Added the Raven: A Poem*. London: George Routledge & Co, 1852.
Schmid, Christian. "The Urbanization of Switzerland." in Neil Brenner (ed.), *Implosions/ Explosions. Towards a Study of Planetary Urbanization*, 268–275. Berlin: Jovis Verlag, 2014.
Schmitt, Carl. *Der Nomos der Erde*. Berlin: Duncker § Humblot, 1974, or. 1950.
Sheerbart, Paul. *Architettura di vetro*. Milano: Adelphi, 1982 (or. title: *Glas-Architektur*, 1914).
Sloterdijk, Peter. *Shaume, Spheren III*. Frankfurt: Shurkamp Verlag, 2005.
Smith, Neil. *American Empire*. Berkeley and Los Angeles-London: University of California Press, 2003.
Soja, Edward W., and Miguel Kanai. "The Urbanization of the World." in Neil Brenner (ed.), *Implosions/Explosions: Towards a Study of Planetary Urbanization*, 142–159. Berlin: Jovis Verlag, 2014.
Taut, Bruno. *La dissoluzione della città*. Faenza: Faenza Editrice, 1976 (or. *The Dissolution of Cities*, 1920).
Taylor, Paul. *Extraordinary Cities*. Cheltenham: Edward Elgar, 2013.
Weil, Simone. *Cahiers, II*. Paris: Librairie Plon, 1972.

6 Luigi Ghirri

The Theater of Landscape

Giuliano Sergio

The transformation of the urban, suburban and rural landscapes of Italy in the 1960s and 1970s was one of the most significant developments in the aftermath of the Second World War. The changing relationship between city and countryside led to a profound transformation of the previous territorial balance. The disruption of this balance between centers and suburbs along new roads of communication marked a profound evolution in the social and political value of the territory. This change affected the major urban 'protagonists' of industrial development and also brought about a capillary diffusion in the relationship between urban and rural landscape, a shift that also concerned the perception and the use of the Italian territory. The symbolic, cultural and anthropological value of landscape is at the center of Italian visual culture. The profound changes during the second half of the 20th century have been in particular the subject of a great deal of attention in Italian cinema and photography.

Nature and *artifice* are the two extremes of an understanding of the reality that has developed in Italy according to a specific dialectic, that overlapped and confused the two elements over the centuries, forging the idea of *landscape*. Landscape is a meeting place where nature always welcomes and preserves the anthropological signs, the artifices due to human activity. However, landscape also has an aesthetic value: it is a pictorial and photographic genre, a way of depicting and representing nature and the relationship that it establishes with man. Landscape in other words is the synthesis of a way of understanding, living and picturing the environment. Even these few elements allow us to understand how this notion is closely linked to the thought of a public artistic activity and the urban and architectural idea of the environment that was developed in the 1970s in Italy, when a radical reflection on aesthetic and anthropological values questioned art and artworks and, at the same time, looked at the environment with a new ecological sensitivity discarding the modern idea of progress brought about by industrialization that dominated in the first half of the century, to consider the landscape as an ecosystem.

The aim of this essay is to follow the transformations of the notions of *nature* and *artifice* in the debate of the artistic culture of the 1970s, through the work of a great figure of the period, the Emilia Romagna-born photographer Luigi Ghirri. As a narrative instrument capable of transforming the vision of Italian territory, photography plays a central role in this period. Photographers were called upon to resolve the connection between a cultural heritage—with its famous sculptures, paintings and monuments idealized, decontextualized and commercialized in the reproductions for tourism—and its context of origin, an urban and rural landscape ignored in the complexity of its post-war transformations. The relationship between *nature* and *artifice* became a search for narrative and visual strategies able to bridge the gap between traditional imagery and the new democratic culture of the post-war period. A specific analysis of Ghirri's visual thinking on the urban landscape, filtered through the urbanistic and artistic positions of the time, will show how in those years the ideal extremes of *nature* and *artifice* engendered new representations of the real, new visual and architectural alphabets, new ways of understanding and building the landscape.

Photography, Museum and Territory: From the Monument to the *Aura*

To gain an understanding of the idea of landscape and its evolution in the Italian visual culture of the 1970s it is crucial to look briefly at the idea of the museum, its functions and especially its relationship with the city, an issue shared by the most diverse positions of the time, bringing together architects, critics, artists and art historians. The museum was the reflection of a historical and conservative culture, now felt as an old-fashioned architectural tool that did not respond to the transformations of contemporary aesthetic culture. Its narrative model continued the idea of a culture separate from the urban context, unable to offer that synthesis between modernity and tradition that characterized the evolution of the Italian cities and landscape. The debate of that period therefore considers the ancient art museum as a "closed opera," a "museum of itself,"[1] a structure that had run out of function in relation to the territory:

> The museum has probably fulfilled its role between 1796 and 1860, that is, in the age of the greatest physical occurrence of conservation, caused by the suppression and deep socio-economic tragedies of the country. The places of original conservation were suppressed, that is, destroyed, sold or modified. The museum was the place where the works could find shelter, a new shelter; As well as conservation, brightness, study and social use.[2]

Since the 1950s, specialists in the field began to refuse the aesthetic conception of the museum and its system of presentation of the artwork inherited from the 19th century. Once its function of hosting and protecting the artistic heritage had become superseded, in the eyes of artists and critics the museum was now only a system of ideological fruition, structured to highlight the formal values of art. Giulio Carlo Argan and many historians of those years looked at the museum as a superstructure replacing the original context of art to transform its presence and value. The apparent "historical immunity,"[3] typical of the museological presentation that allows the comparison between works of different epochs and places, was a cultural construction questioned in those years.

> The museum is the place where the artwork is deprived of any remaining remnant of its historical existence and presented in its pure, permanent, absolute artistic dimension—Giulio Carlo Argan wrote in a famous essay published in 1968—applying Benjamin's terminology, we can say that in the museum the artwork loses all its *aura* and, in return, achieves the maximum degree of *exhibition* [. . .]: what is expressed as absolute and immutable value is in fact only the value that the work of art has in the present, that is not in and of itself, but in our cultural context. Therefore, the architecture of the museum is, it must be architecture.[4]

The rooms of museums make up a great scenic machine that, with its self-proclaimed historical objectivity, updates the artwork and transform its aesthetic function. The fact that Argan had to turn to the Benjaminian categories of *aura* and *exhibition*—used by the German philosopher to talk about reproduction photography—clearly shows the aesthetic and conceptual link between the museum presentation and the photographic representation, their common nature as *narrative machines* of art. Walter Benjamin, in his famous essay *The Work of Art in the Age of its Technological Reproducibility*,[5] revealed that the disappearance of the *aura* in the photographic reproduction of works was the result of a photographic style born in the second half of the 19th century to satisfy the encyclopedic and museological concept of art history. This reproduction of the artwork and its details exalted the formal and structural aspects of the artistic object, eliminating its context, its *aura*, and allowing the comparison of works outside their original anthropological context. The disappearance of the *aura* was therefore the effect of a precise kind of photography, with its own style, capable of realizing a visual museum of the entire artistic heritage. Benjamin's essay indirectly announced a crucial chapter for cultural strategies that would later develop in the post-war period, when the ability to reproduce and present works of art to the masses coincided with the growing need of administrators to look at art as a cultural asset to promote a didactic, democratic and identity enhancing role for the national heritage. This cultural policy that will have in the exaltation of the museum and in the diffusion of heritage through photography the main tools of a stylistic and formal narrative, had its theoretical manifesto in André Malraux's *Musée Imaginaire*.[6]

These few remarks on the modalities of representation of art should not be seen as outside the scope of this essay. A consideration on the relationship between nature and artifact at a precise historical moment must be understood as an analysis of narrative dynamics prevailing in a particular culture. Their study allows us to measure the positions that prevail in the artistic sector within a specific historical period. In Italy, cultural heritage is rooted in landscape, which is why studying the museum debate in this period helps us understand the city and its relationship with its outskirts as well as the redefinition of the landscape idea as an artistic and photographic genre.

As we have seen, in this period, the museum was perceived as a place of cultural and patrimonial alienation, using an obsolete model that neither met the needs of artists nor reflected the culture of the territory. The strongly regionalized social and political tradition of Italy led to a specific assessment of the museum system and its relationship with the territory—a "natural museum" as it was called by André Chastel[7]—that would have led to a different consideration of photographic work and its aesthetic function. Distances have been taken with the post-reunification tradition of the great 19th-century photographic studios like *Brogi* and *Alinari*, who had built the official and tourist vision of the Italian landscape and its artistic heritage.[8] Scholars like Andrea Emiliani, Superintendent and Director of the Pinacoteca Nazionale in Bologna, questioned the traditional concept of the museum and the iconography it was championing, supporting the awareness of the territory as an original place, an irreplaceable *aura* of the artistic object and the monument:

> The traditional concept of conservation of artwork as a self-evident entity has now shown its inadequacy. [. . .] A new concept of conservation as a public service capable of addressing the problem of the protection of the Italian cultural heritage in its entirety: a heritage to be protected by protecting not only the works of

art, but also by saving nature and the landscape, guaranteeing the historic centres retain their integrity, protecting lesser monuments, recording local traditions.[9]

This attention to the 'local' will lead to a renewal of monitoring activities over the territory and the need for a photography that not only reproduces objects and palaces, but knows how to capture the *aura* of the places, their urban, landscaping, cultural and anthropic context. A sensitivity that required a different photographic approach, capable of breaking with a tradition only interested in major monuments.

The Landscape of Luigi Ghirri, Between Architecture and Avant-Garde

The techniques of representation of the landscape and the use of photography are a privileged place to analyse the influences and collaborations between artists, architects and photographers. The debate on the representation of the city and the territory had already animated the photographic research of the historical avant-gardes, interested in the New Vision, a tendency that in Italy often came through the experimentation of architects and the attention of magazines such as Domus and Casabella.[10] After World War II, the rejection of academic and museum hierarchies promoted by the neo-avant-garde, coincided with an urban sensibility that advocated abandoning the canons of the 19th-century landscape painting[11] and looked at the city and the transformation of its relationship with the countryside. An anti-monumental vision of the territory and art emerged that in turn led to a common way of reading the urban context, considered as a unique ecosystem to be explored in its complexity. Some architects turned with interest to the installations of the neo-avant-garde, able to configure the city spaces differently.[12] The artists seemed able to rethink the categories of architecture and sculpture in a direct confrontation between the monumental legacy of the past and the popular and industrial culture of the time.[13]

Luigi Ghirri started his career in the late 1960s, collaborating with artist friends from Modena in the documentation of installations and works and experimenting with new narrative possibilities in the use of photography. His experimentation gradually developed during a period of radical critique of traditional values, in which it was difficult to define the function and role of photographers in Italy. Ghirri accepted the abolition of the aesthetics proposed by the avant-garde: documentary practice and the rejection of the reportage was a fundamental starting point that allowed him to grasp what was happening between art and photography. His visual language appropriates the aesthetic stimuli that pervaded the artistic culture: from pop art to conceptualism, their surreal and metaphysical roots, all the way to paradox and visual alliteration. This iconographic culture became the basis on which to build his own vision of the Italian landscape. The city, its suburbs and the contrasts that popular architecture expresses, are some of the main components of Ghirri's work. He produced a collection of images which depict the transformation of Italian space and sites, and composed a visual puzzle out of a new urban imagery which emerged during this period. The dual approach chosen by the photographer for an interpretation capable of overcoming the academicism of the Italian photographic vision was irony and poetry (Figure 6.1). For Ghirri, it was a way to dismantle the consolatory vision of the landscape, the iconographic harmony of an agreement that in everyday urban reality was replaced by the visual paradox, contradiction and mockery of the loss of every aesthetic reference.[14]

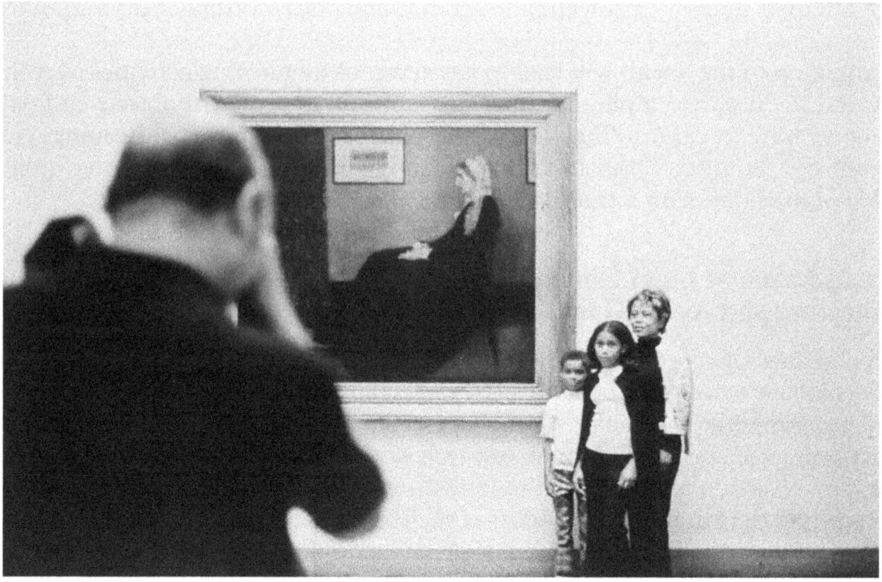

Figure 6.1 Luigi Ghirri, *Parigi*, 1972 (original photograph in color)
© Eredi Luigi Ghirri

In the 1970s, Ghirri discovered American documentary photography, which was launched in the 1960s by the MoMA in New York, and was showcased in Italy at the Parma CSAC.[15] Walker Evans was the artist closest to Ghirri's sensibilities and his critical vision of photography: "Evans is the author that I loved, love and respect more than anyone else, and I feel closest to. I first saw Evans' photographs in 1975, and I think they were important for my work, for what I had done and was doing and for its further development."[16] In the United States, Evans was the photographer closest to the first avant-garde, the recognized master of the *New Documents* photographers. Even pop and conceptual artists had turned to his work, adopting the language of his photography, which explored the boundaries between document and vernacular photography.[17] With its notion of documentary style, Evans had opened up the possibility of a photography that developed a second level of interpretation, building the representation of reality with a critical filter capable of neutralizing the rhetoric of traditional photographic genres. Evans' position coincided with Ghirri's project of deconstructing styles and narratives that had come before him to establish a new photographic language capable of responding to the stratifications of Italian culture.

Paesaggi Di Cartone, or the Possibility of a New Landscape

At the end of 1973 Luigi Ghirri published his first portfolio, *Paesaggi di cartone* (Cardboard landscapes).[18] This work appears today as the mature announcement of a language that will lead the photographer to the neutral style of his later urban landscapes. Ghirri effortlessly rejected the dramatic subjects of reportage and the formalisms of subjective photography, proposing a vision that takes photography itself and its presence in the urban environment as its subject. The introduction Ghirri wrote for *Paesaggi di cartone* offers many insights into his photographic and artistic approach:

> When I travel, I take two kinds of photographs, the usual ones that everybody takes, which ultimately are of little or no interest to me, and then the others, the ones I really care about, the only ones I really think of. In my "pictures" the subjects are those we see every day, they belong to our usual visual field: they are images that we are accustomed to passively enjoy; Isolated from the usual context of the surrounding reality, proposed again in a different mode, these images reveal themselves charged with a new meaning.[19]

His photographs capture billboards, shop windows, signs, elements of a landscape that shows itself only through the mediation of the icons that inhabit it. Ghirri seems to want to limit himself to photographing the popular and ephemeral imagery invading the landscape,[20] trying to submit these obsessive presences to a critical and poetic gaze. What is announced with *Paesaggi di cartone* is therefore a deep break with the idea of the unitary representation of the city within its urban landscape: what emerges is a fragmented territory, marked by a superposition of signs. "The sense I try to give to my work is to verify how it is still possible to desire and to face the path of knowledge in order to finally distinguish the precise identity of man, of things, of life, from the image of man, of things, of life."[21] Ghirri's landscapes are graced with a calmness, which avoids the interpretation of urban space, to favor its description through popular culture, details of shop windows and trompe-l'oeil images. Through the collection of these elements, Ghirri's work creates an apparently anonymous gaze that, by recapturing the documentary style, allows him freedom from traditional genres. Thanks to this construction of the image, Ghirri comes to obtain a temporal suspension that often imposes an almost metaphysical spatiality. His works construct a system of logical and formal references that give the impression of observing a "thought through image," an idea evoked several times in his texts as a central concept of the photographer's poetics (Figure 6.2).

> The possibility of the analysis in time and space of the signs that make up reality, whose entirety always eludes us, thanks to fragmentation, allows photography to be closer to the non-definable, namely to physical existence. This is the reason why I'm not interest in: the decisive moments, the study or analysis of the language as an end in itself, the aesthetics, the totalizing concept or idea, the poet's emotion, the erudite quotation, the search for a new aesthetic belief, the use of a style.[22]

Landscape as Narration

Ghirri laid the foundations for a "new visual alphabet"[23] that offers an alternative to traditional photographic genres. The development of a photographic language brought Ghirri to develop the elements of a different narrative: "I believe that storytelling still exists. In this regard, I think it is impossible to escape the narration, even when it seems to be completely denied."[24] The image is an element that must be read within a context, in relation with the other photographs in the series. Ghirri considered his pictures 'maps' that at the same time are also photographs. Their meaning is

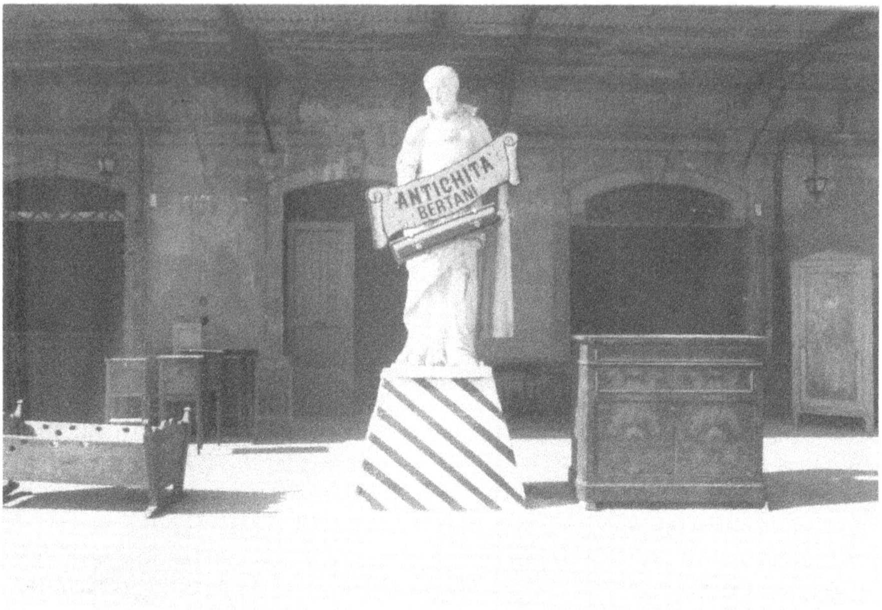

Figure 6.2 Luigi Ghirri, *Antichita Bertani* (original photograph in color)

not that of direct reference to reality, but an immersion into visual language, into the hieroglyph: this allows the photographer to find new ties and relationships between his photographs. His series of the 1970s were always open to new configurations, images were added over the years, making a continuous reconfiguration of his archive.

For Ghirri the archive must not be understood as a progressive cataloging, as a way to preserve a serial reproduction of every aspect of reality,[25] but as *the image of the images*, the atlas, a space where everything is contained, or even as a theater of thoughts, a place where the author's works are collected, placed together and referenced, where the author can builds a map of references, a visual itinerary, a mental reflection on the landscapes from which they originate. So, for Ghirri, images are signs, personal places: "These places are also my place, my room: on the walls the figures: landscapes and portraits of the ancestors impose me a detachment that does not forget a thin adhesion."[26] The archive becomes the space of a possible narrativeness of the image that includes the iconographic tradition that preceded it, where the relationship between nature and artifice, once the ancient harmony is lost, finds new configurations, associations and assonances and touches the roots of language, its possibilities of articulation, giving rise to a new form of poetry.

Landscape as a Choice

The terms used to describe Ghirri's work—*archive* and *theatre*—demonstrate its conceptual connections with the museum—the collection and the possibilities of its display—disclosing their common link to the central question of narration. Compared to traditional culture, a new vision was progressively imposing an aesthetic sensibility that did not separate the artistic heritage from the contemporary landscape, which offers a continuous reading that exceeds the academic divisions embodied by the architecture of the museum.

The avant-garde artists were the first to develop the aesthetics of the archive and the possibilities of displaying it by moving it from the private workplace to a new form of representation of the artist's work-life. Even in Ghirri, the archive, intended as a theater, is a linguistic tool and a mode of thought, and allows a distance from the shallows of intimist narrative. One such example is the *Identikit* series (1976–1979) where the photographer had identified the interior of his home as a privileged place, a lounge-museum where to find the signs of an intellectual biography. "I have delegated for this self-portrait the objects (books, records, etc.) that testify of a relationship of knowledge, of culture, of my imagination, of spending my time."[27] The *Identikit* series consist of frontal images of his own library that testify to the omnivore knowledge in which the author disperses and finds himself in his shots. This work already has all the characteristics of the new approach and shows how the iconographic culture of the avant-garde, with its attention to the context of everyday life and the surrounding environment, is part of the cultural background of the photographer. Is a minimal aesthetic that we will find used as a key to access the urban, vernacular world of the province where he lives.

Ghirri's attention to architecture precedes his passion for photography, because it is based on his profession as a building surveyor. For him, architecture is the background and backbone of existence and the building elements are a fundamental tool of his photographic narration, becoming symbols and mirrors of his idea of landscape. In these years, Ghirri theorized his idea of a minimum photographic journey across the new cityscape: "This is why I am particularly interested in the urban landscape, the suburbs, because it is the reality that I have to live with every day, the one I know best and that I can therefore best reproduce as a "new landscape" for a continuous and systematic critical analysis."[28] In his first architectural survey of Modena, Ghirri explores the materials, forms and structures of the local architecture: he looks at the architectural details of public housing in his neighborhood (*Catalogo*, 1971–1973). His architectural reading leads him to evolve a photographic vision that analyzes and isolates the structural elements of buildings, establishing a poetic alphabet that reflects the popular culture of the Italian suburbs.

> I had an accentuated interest in certain places that we can call architecture. The houses that made up the street I lived in, the roads I walked every day, were and are architectures. The doors, the colours of the plaster, the cladding, the vases that decorated the windows were the tiles of the buildings, the tiles that I investigated with affection. Just because anonymous and lost, these architectures seemed to wait for someone to give them an identity. What I did between 1970 and 1975, photographing the edges of ancient cities, the suburbs, or predominantly those villages without historical or geographical dignity, was a kind of re-composition of a family's album of my as well as our external reality.[29]

In this photographic taxonomy of popular architecture of the 1970s, he soon introduces the presence of nature: in his vision the nature tamed and forced reveal herself in simple architectural and decorative forms, or growing in the interstices of the courtyards, in pots and in flower beds and in environments which seem to have forgotten any previous agreement with nature. The series *Colazione sull'erba* (1973) is composed of a melancholic herbarium, revealing the naive imagination of those who inhabit the modern city structures, an imagery that, without knowing it, has lost its ancient relationship with the countryside. The opposition within nature and artifice became an ironical myth, an iconographical stereotype, an allegory, that allow us to query the landscape as a mirror of society facilities, where to measure the popular perception of nature, industrial culture and artistic heritage. This structure of Ghirri's vision helped him compose projects that let him advance a reconsideration of the landscape in which to explore the new relation of nature and artifice in popular imagination. In works such as *Italia Ailati* (1971–1979), *Il paese dei balocchi* (1973–1977), *In scala* (1977), the depiction of nature and artifice becomes an investigation on the representation of that relation. Ghirri always mentions the subtle limits between the authentic and its copy, between the landscapes and their disguises. He conceives his work as a denunciation of mass culture, which overlaps and erases reality: "the recovery of experience [must] necessarily be effected starting from the awareness of destruction."[30] His awareness accompanies the photographic work from the beginning and poses new problems in his relationship with reality, "the possibility of finding pockets of transparency, or, rather, finding the centres, is becoming more and more difficult. The targets move at an ever-increasing speed. The problems of recovering visions 'on man' or 'of man' move on too familiar paths [. . .] today it is difficult to photograph."[31]

The landscape in Italy is the main challenge that the artist faces. It is a genre where every city, every rural area and region has its own pictorial and photographic iconography to relate to.

> Selecting an Italian imagery is much more difficult than in other countries such as the United States. The visual imagery of the twentieth century is, in our country, composed of images derived from photographs by Alinari, school textbooks, illustrated postcards, Touring Club books and a thousand other places [. . .] this tends to build what I would call the "common place"—wrote Luigi Ghirri—The problem is to see, through all the previous images, that common place and, at the same time, to erase it in order to find our own first vision.[32]

The elements of architecture and the *paesaggi di cartone* gathered by Ghirri in his works of the 1970s are elements that, in the landscapes of 1980s, were structured into a theater of things: places that, through silence, testify to the signs of the community. The theater is perhaps the easiest way to approach Luigi Ghiri's work.[33] Here the theater is not to be understood in its specific notion, linked to scenic action or architectural typology, but rather as an expression of a wider idea of place and community—the theater of human affairs—the city and the landscape. Photography and architecture are similar languages, seen through the example of the museum both can build a complementary narrative, a point of views that Ghiri carefully avoids: for him to photograph does not mean affirming his own style, adding new images or constructions, but offering a vision where the monuments, landscapes and anonymous architectures of the community that expressed them coexist in a single unit presented in and respected by the image. In the 1980s the photographer's gaze extends, looking at certain places with a different focus, building ties with their history, in a ubiquitous cultural anthropology, to an ever-humanized nature. This is the evolution of the sensibilities of a man who coincided with the new season of the artistic and photographic scene. Ghirri's photographic series *Il profilo delle nuvole* (1989) was an organic project meant to depict the landscape as a mirror of the community. An example is a photo from 1986, taken in Ravenna, in which the dusky blue of the sky envelops the set design Aldo Rossi created for Donizetti's Lucia di Lamermoor (Figure 6.3). Ghirri's rapturous gaze captures this square within a square, where fake towers, palaces and bell towers are superimposed onto real buildings. The architectural double is revealed by a small illuminated window in the top right-hand corner, which shows the mutable border between the set and the city-theatre which houses it. The domestic light intruding on the scene is the sign of continuity Rossi was striving for and is revealed by the photographer as the limit of a tautology where the machine ends and the city begins. This is what the architecture becomes precisely when it questions and crosses that border, when it projects a place as the mirror of a renewed experience of the urban landscape. The book *Il profilo delle nuvole* retains the idea of a sequence that makes up emotional juxtapositions, overcoming geographies, itineraries and chronologies: "If photography is a journey it is not in the classical terms suggested by this word, it is rather a path with many bends and U-turns, randomness and improvisation, a zigzag line."[34] It shows a succession of landscapes and interiors, of churches and theaters, frescoes, squares and gates in the countryside. The attention to photographic genres joins the rediscovery of an iconographic tradition and its roots: a reconfiguration of history as the history of meaning, as reactivation of sediment accumulated over time. Ghirri transforms the landscape genre: photography has to look beyond its functionality to reach a dialog "that allows a face, a place, a landscape to be recognizable, familiar, habitable."[35] The images of the series

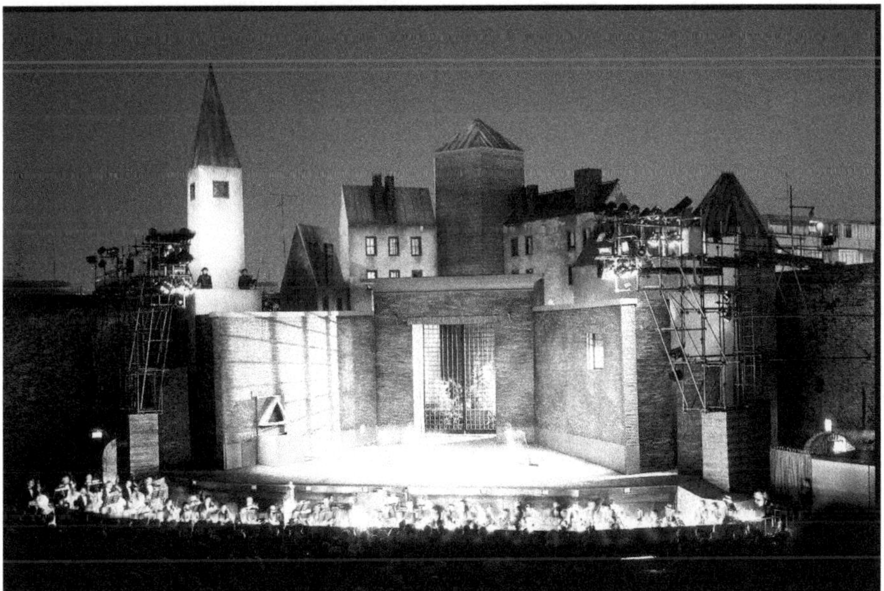

Figure 6.3 Luigi Ghirri, *Ravenna*—Scenografia di Aldo Rossi per la Lucia di Lammermoor, 1987 (original photograph in color)

establish a complex relationship between the representation, the real and the public who observes, bringing into play the viewer and the place he occupies in front of the work: an overturned image structure where the representation brings us back to the gaze of its author. At every photo Ghirri affirms the possibility of *continuing to see the world*, to be able to make a judgment, to be able to approach beyond his monotonous appearance, without thereby structuring a style. That partly explains the enigma of his vision: Ghirri judges outside of a pre-established discourse, the thought comes from the real, in the form of image, and brings us to the author and to the expectation of the next image.

Ghirri's photographs, contrary to the reproductions Benjamin was talking about, capture the *aura*, it is as if the museum lost its walls, becoming an inconsistent monument, a frame that directs the eye towards the territory, which allows us to rediscover the *artworks* of man between palaces and squares, between the glimpses of villages and the countryside, seizing them in the folds of history. Ghirri's photographs reveal theaters invisible to our eyes, eyes that have become too accustomed to their presence: they are images that play with the commercial and awkward overlays of vernacular architecture, that are transfixed in front of the ancient stories narrated by frescoes, places that the photographer finds crowded with stage props and sets accumulated over time. The landscape of the plain and its buildings then becomes a moment of attention, the awareness that the *natural* proportions of those spaces are the result of an ancient sedimentation, a common heritage that must be understood and interpreted. It is no coincidence that Ghirri's works look to the province: if his narrative is not intimate, if his gaze is not only literary but has successfully navigated the pop and conceptual season, if his education is architectural and historical-artistic, then the choice of photographing the province, its ingenious contradictions, its urban transformations that exchange unaware dialogs with the treasures spread throughout the territory, is a deeply political choice. Politics is the choice of style, where color photography offers a neutral, documentary view that avoids screaming and militant denunciation of black and white. A choice that does not criticize the new suburbia of the big cities that sprung up in the 1970s but simply excludes them,[36] aware of composing a reading of society, a vision that narrates an epoch, that offers a wealth of images that are the perfect vantage point from which we can review the past season and rethink our future.

Notes

1 Cfr. Andrea Emiliani, *Il museo, opera chiusa*, in Andrea Emiliani, *Dal museo al territorio* (Bologna: Alfaeditoriale, 1974), 51.
2 Ibid. Author's translation, and thereafter unless indicated otherwise.
3 Giulio Carlo Argan, *Il Museo d'arte moderna*, in *Metro* (14 June 1968), 5.
4 Ibid.
5 Walter Benjamin, *The Work of Art in the Age of Its Technological Reproducibility, and Other Writings on Media* (Cambridge: Harvard University Press, 2008, original ed. *Das kunstwerk im Zeitalter seiner technischen Reproduzierbarkeit*, Frankfurt 1955).
6 The Post-Benjaminian reflection will lead to new approaches for the promotion and study of artistic heritage. In France, André Malraux's *Musée imaginaire* (Geneva: Skira, 1947) will be the most lucid formulation of a vision that accepted and brought to the extreme consequences the eradication of artworks made by museums and amplified by photographic reproduction. Malraux's perspective shared Benjamin's considerations on the multiplication of the work done by photography to reach a history of art that is structured in the virtuality

of reproductions. Thanks to reproductions, illustrated books and great shows, it was possible for Malraux to overcome the physical and historical *aura* of art objects and give them a new dimension by building the most diverse paths in the virtually infinite and *complete* space of the imaginary museum.

7 In his text *L'Italia, museo dei musei,* André Chastel writes: «In the end, the peninsula must be regarded as the place par excellence of the "natural museum" »; he unequivocally clarifies the cultural extraneousness of the Italian tradition to the vision proposed by Malraux, speaking explicitly of « uselessness or absence of the *Musée Imaginaire* »; in Andrea Emiliani (ed.), *Capire l'Italia : I musei* (Milan: Touring Club Italiano, 1980), 12–14.

8 For an analysis of the models of city representation in Alinari's photographic campaigns see Carlo Arturo Quintavalle, *Gli Alinari* (Firenze: Edizioni Alinari, 2003).

9 Andrea Emiliani, *La conservazione del patrimonio culturale*, in *Rivista IBM*, vol. VIII (1972), 2, 38.

10 These architecture magazines were valuable sources of information on international research during the Fascist period. See the introduction by Cesare De Seta to *Giuseppe Pagano fotografo*, exhibition catalog edited by Cesare De Seta, (Rome, Istituto Nazionale per la grafica), (Electa, Milan 1979), 5–12.

11 Cfr. Cesare De Seta ed., *Paesaggio, Annali, Storia d'Italia*, vol. 5, (Torino 1982).

12 Some architectural firms such as Archigram in England or Archizoom and Superstudio in Italy have developed a concept of the project with urban visions that uses pop visual culture and photomontage. At the basis of this hybridization of languages, there was the belief that a different technique for the vision of urban reality would have allowed a different approach to the realization of projects free from the modernist spatial conception. See Roberto Gargiani and Beatrice Lampariello, *Superstudio* (Rome/Bari: Laterza, 2010) and Roberto Gargiani, *Archizoom Associati 1966–1974. Dall'onda pop alla superficie neutra* (Milan: Electa, 2007).

13 For a broad analysis of this period and the relationship between artists and photographers, I allow myself to refer to my book: Giuliano Sergio, *Information, Document, œuvre. Parcours de la photographie en Italie dans les années soixante et soixante-dix* (Paris: Presse Universitaire de Paris Ouest, 2015).

14 A paradox that in Italy was first presented by film-makers through the exploration of the suburbs whose sense of displacement and estrangement that Federico Fellini, Pier Paolo Pasolini, Marco Ferreri and Michelangelo Antonioni used to define their characters.

15 In 1971, the University of Parma, through the Centro Studi e Archivio della Comunicazione (CSAC), developed an intense program of exhibitions on American photography in order to promote a new photographic culture in Italy. The first was New Photography USA (1971), with works by Diane Arbus, Paul Caponigro, Jerry Uelsmann, Joel Meyerowitz, George Krause and Bruce Davidson. Others followed, Dorothea Lange (1972), a retrospective about the Farm Security Administration (1975) and a personal exhibition by Lee Friedlander (1975).

16 In Luigi Ghirri, *Vedute (1970–1979)*, in Luigi Ghirri, *Luigi Ghirri*, exhibition catalog (Parma: CSAC 1979) 81.

17 Peter Galassi, *Walker Evans & Company* (New York: The Museum of modern art, 2000), Jean-François Chevrier Allan Sekula, Benjamin H.D. Buchloh, *Walker Evans & Dan Graham* (Rotterdam: Witte de With, 1992) and Clément Chéroux, *Walker Evans*, (San Francisco: The Museum of modern art, 2017).

18 Cf. Luigi Ghirri, *Paesaggi di cartone,* in *Il diaframma. Fotografia italiana*, n°188, (December 1973), 28–32.

19 In Luigi Ghirri, *Paesaggi di cartone* , op. cit., 28.

20 Cfr. Luigi Ghirri, *Luigi Ghirri,* op. cit., and Germano Celant, *Bello qui, non è vero?*, (Roma: Contrasto, 2008).

21 In Luigi Ghirri, *Kodachrome* (Modena: Punto e Virgola, 1978), 12.

22 Ivi.

23 Cf. Luigi Ghirri, *Niente di antico sotto il sole*, (1988), in Paolo Costantini, Giovanni Chiramonte (ed.), *Luigi Ghirri Niente di antico sotto il sole. Scritti ed immagini per un'autobiografia* (Torino: SEI, 1997), 135.

24 In Luigi Ghirri, "*Vedute*," in Luigi Ghirri, *Luigi Ghirri*, op. cit., 80.

25 For Ghirri "the proliferation of cataloging, [. . .] seems to actually remind us that we are preparing the documents to be taken in the Second Ark of Noah; But in these deposits of objects, gestures, people, I do not consider it to be a wider testimonial validity, [. . .] but a total acceptance of the tic which we would like to deny: collecting as anesthesia of the eye" Luigi Ghirri, *Catalogo*, in Luigi Ghirri, *Luigi Ghirri*, op. cit., 70.
26 Luigi Ghirri, *Introduzione in Paesaggio : immagine e realtà* (1981), in Paolo Costantini, Giovanni Chiramonte (ed.), *Luigi Ghirri Niente di antico sotto il sole*., op. cit., 42.
27 Luigi Ghirri, *Identikit (1976–1979)*, in Luigi Ghirri, *Luigi Ghirri,* cit., 84–85.
28 Luigi Ghirri, *Paesaggi di cartone*, op. cit., 28.
29 Luigi Ghirri interviewed by Mario Lupano, *Fotografare i luoghi, fotografare le architetture*, in Paolo Costantini, Giovanni Chiaramonte (ed.), *Luigi Ghirri Niente di antico sotto il sole*, cit., 294.
30 Luigi Ghirri, Kodachrome, op. cit., 66.
31 Luigi Ghirri, *Re Mida nel vicolo cieco*, *Piero Castellano/Piero Deluca,* catalog of the exhibition, Rimini 1982, in Paolo Costantini, Giovanni Chiaramonte (ed.), *Luigi Ghirri Niente di antico sotto il sole*, op. cit., 44.
32 Luigi Ghirri interviewed by Marco Belpoliti, *Nel regno dell'analogo*, in *Il Manifesto*, (16 March 1984), 7.
33 A notion which Gianni Celati utilized in his text *Commenti per un teatro naturale delle immagini*, written for Luigi Ghirri's book *Il profilo delle nuvole* (Milan: Feltrinelli, 1989).
34 Luigi Ghirri interviewed by Emanuela Teatini, *Un canto della terra*, in Paolo Costantini, Giovanni Chiaramonte (ed.), *Luigi Ghirri Niente di antico sotto il sole*, cit., 297.
35 Luigi Ghirri *Fotografia e rappresentazione dell'esterno*, in Paolo Costantini, Giovanni Chiaramonte (ed.), *Luigi Ghirri Niente di antico sotto il sole*, cit., 83.
36 Ghirri took photos of cities such as Milan, Naples, Rome, Paris and New York. This list could go on, but his depiction of big cities is almost always linked to a defined urban landscape, to a city-theater (theatre?), such as villas and archeological (archaeological?) parks, or the details and traces of urban history. In general it is difficult to find wide vistas of modern, post-war neighborhoods, whereas images of Emilia and Capri landscapes return frequently, as do the seaside towns in Puglia or the famous images of the Palace of Versailles.

References

Argan, Giulio Carlo. "Il Museo d'arte moderna." *Metro*, June 14, 1968.

Belpoliti, Marco. "Nel regno dell'analogo." *Il Manifesto*, marzo 16, 1984.

Benjamin, Walter. *The Work of Art in the Age of Its Technological Reproducibility, and Other Writings on Media*. Cambridge, MA: Harvard University Press, 2008.

Chéroux Clément, *Walker Evans*, exhibition catalogue. San Francisco: The Museum of Moden Art, 2017.

Chevrier, Jean-François, Allan Sekula, and Benjamin H.D. Buchloh. *Walker Evans & Dan Graham,* exhibition catalogue. Rotterdam: Witte de With, 1992.

Costantini, Paolo, and Giovanni Chiramonte (ed.). *Luigi Ghirri Niente di antico sotto il sole. Scritti ed immagini per un'autobiografia*. Torino: SEI, 1997.

De Seta, Cesare (ed.). *Giuseppe Pagano fotografo*, exhibition catalogue. Rome, Istituto Nazionale per la grafica. Milan: Electa, 1979.

De Seta, Cesare (ed.). *Paesaggio, Annali, Storia d'Italia*, Vol. 5. Torino: Einaudi, 1982.

Emiliani, Andrea. "La conservazione del patrimonio culturale." *Rivista IBM* VIII (1972): 2.

Emiliani, Andrea. *Dal museo al territorio*. Bologna: Alfaeditoriale, 1974.

Emiliani, Andrea (ed.). *Capire l'Italia: I musei*. Milan: Touring Club Italiano, 1980.

Galassi, Peter. *Walker Evans & Company*, exhibition catalogue. New York: The Museum of Modern Art, 2000.

Gargiani, Roberto. *Archizoom Associati 1966–1974. Dall'onda pop alla superficie neutra*. Milan: Electa, 2007.

Gargiani, Roberto, and Beatrice Lampariello. *Superstudio*. Rome/Bari: Laterza, 2010.

Ghirri, Luigi. "Paesaggi di cartone." *Il diaframma*. Fotografia italiana, n°188, December 1973, 28–32.
Ghirri, Luigi. *Kodachrome*. Modena: Punto e Virgola, 1978.
Ghirri, Luigi. *Luigi Ghirri*, exhibition catalogue. Parma: Quaderni / Università di Parma, Centro studi e archivio della comunicazione, 1979.
Ghirri, Luigi. *Il profilo delle nuvole*. Milan: Feltrinelli, 1989.
Malraux, André. *Musée imaginaire*. Geneva: Skira, 1947.
Quintavalle, Arturo Carlo. *Gli Alinari*. Firenze: Edizioni Alinari, 2003.
Sergio, Giuliano. *Information, Document, œuvre. Parcours de la photographie en Italie dans les années soixante et soixante-dix*. Paris: Presse Universitaire de Paris Ouest, 2015.

7 Nature, Plastic, Artifice

In Conversation With Tuula Närhinen

Tuula Närhinen and Gabriel N. Gee

Tuula Närhinen is a visual artist based in Helsinki, Finland, whose artistic practice mirrors and appropriates scientific experimentation methods and history to develop a poetic reflection on the present metamorphoses of natural entities and environments. A particular interest in 19th-century scientific investigations on nature, in which the scientific goals were closely intertwined with aesthetic qualities informs her work. She has produced organic photographs (*Solarium*, 2004), evoking the early days of photography when Henry Fox Talbot could name his chemical achievements 'the pencil of nature' (1844–1846), used chromatography to separate pigments in the color of flowers (*Chromatograms*, 2002), and made photograms of wavescapes (*Clapotis*, 2009). The tracing of the wind on paper (*Windtracers*, 2000), of the movement of animals who share our environments (*Tracing Animals*, 2005), and of water currents through a bottle equipped with a light (*Surf*, 2001), are all means to represent the surrounding presence of nature. Närhinen in that respect echoes the work of the biologist Ernst Haeckel, who coined the term 'ecology,' and resorted in his professional practice throughout his lifetime to accomplished drawing skills. Through her studio located on a small island out of Helsinki, her work has been particularly drawn to the sea. The series of sculptural work she has made out of plastic found on the seashore, furthers a research on the impact of human global industrial agency in the 20th century, and the ambiguous beauty that can be found in the all-pervasive outreach of plastic waste on maritime environments. In the following discussion, Närhinen starts by evoking her sculptural series *Baltic Sea Plastique* (2013), which stems from her interest into plastic pollution. The conversation evokes evolutionary theories, together with the formation of hybrids that our society of consumption taking off in the post-war era has fomented. The dialog moves into a reflection on the parallels between science and the arts, and the construction of reality that these two domains of human activity share in common, beyond the standard division between 'fact' and 'fiction.' The notion of 'artifice' in artistic and scientific production is finally evoked, in considering both the production and modes of presentation of the work.[1]

Gabriel N. Gee: You are working from a little island off the coast of Helsinki, where you have your studio. A lot of your work revolves around this location. Can you tell me about the significance of this space?

Tuula Narhinen: I started working on this island 15 years ago. The city of Helsinki owns a house here where they rent studios to artists. There are 25 artists currently working on the island in this house, with different practices; there are some writers, some visual artists like me.

The building itself is very special because it was built for the Finnish army to do chemical experiments. It's not a normal studio space, it has very old rooms, with particular interiors that cannot be changed. For sculpture it can be difficult, as you need to cross the sea every time you go to the studio from Helsinki, and that's one of the reasons I work with the sea so much. There is a connection boat, but it's only during summer time. In winter time we need to row or walk on the ice to get there! There are times I can't go, as the journey finds itself between ice and the open sea beyond. But it gives me a lot of inspiration, because every time I go I practically touch the sea. I get to observe the state of the sea, and so I've done a lot of work with marine pollution and plastic.

G.G: One of your sculpture series is entitled *Baltic Sea Plastique*. It consists of a number of sculptures that you make out of plastic waste found in the sea. It's quite a recent piece (2013), how did you initiate this work and this reflection on plastic errance? (Figure 7.1)

T.N: I started the project a few years ago. I have a long term research interest into plastic pollution. I did two other works prior to this one about plastic waste, one of which is called *Mermaid's Tears* (2007), which is a necklace made out of micro plastic, while the other, *Frutti di Mare* (2008), is made with bigger plastic detritus. Together they are called *The Plastic Ocean*. The plastic theme is an important component of my research interest. These works were in the background when I started *Baltic Sea Plastique*. With *Frutti di Mare*, I had created sculptures that were placed in an aquarium; then I experimented with the

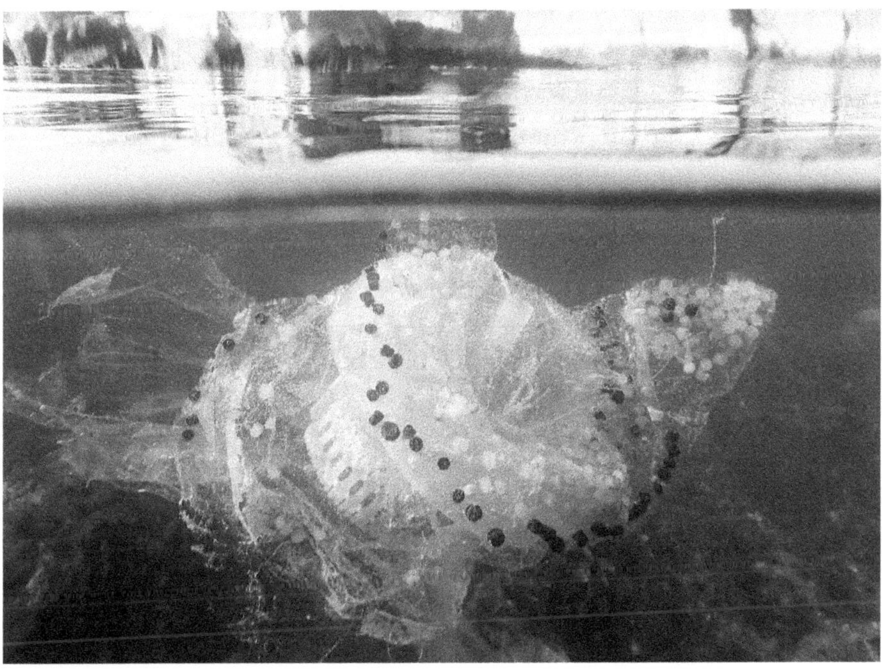

Figure 7.1 Tuula Närhinen, *Baltic Sea Plastique*, 2013
courtesy of the artist

floating, how the water carries some parts of the objects, while other parts sink. With this method and the material used I aimed to claim a part of the Helsinki seashores. Not all the plastic components come from the island where I work, as it's a fairly small island. But I take my bike and I go around the Helsinki seashores, with a big black plastic bag in which I collect trash, which I then bring to the studio to sort it out, according to colours, or textures, whatever inspires me. In *Baltic Sea Plastique*, I was interested in how the plastic behaves and moves in a way that also captures the attention of sea birds and other animals living in the sea. Plastic is a very strange material, because it imitates the appearance of the jellyfish and other creatures which are part of the sea, which is very unfortunate because then the plastic parts end up in the intestines of birds. One of the points to underline about plastic is that it is a man-made thing which is now really out of our hands. There is a cycle of design, as it's a very polyvalent material, it can take any shape, it's a dream material for designers, you can make it thin and strong, and any texture is possible. It's designed very carefully, it is then consumed, and people then throw it away . . . It leaks into nature, and then it comes back as trash, and finally an artist puts the bits and pieces together! Of course from the biological point of view, from the nature point of view, the story is really sad, as the material is deceitful.

G.G: In your accompanying text to *Baltic Sea Plastique*, while there are allusions to the dramatic aspect of this phenomenon, and to the ecological implications that stem from it, the stress is more significantly on metamorphosis, and processes of historical change that are channelled by the material.

T.N: Correct, I have written more about this aspect in Finnish, as part of my practice-based doctoral thesis at the University of the Arts in Helsinki. In a chapter dedicated to this work, I go as far as framing this metamorphosis as an evolutionary process, and refer to the German 19th-century biologist Ernst Haeckel (1834–1919), who made research on very small marine organisms. Haeckel also reflected on the manner through which everything is interconnected in the world. He was a marine biologist. I made a connection with oil, which is the base of plastic, and some of the organisms he was looking at that belong to the same layers in the bottom of the sea. Haeckel made very interesting illustrations related to his research, in particular in *Kunstformen der Natur*, one of his books, and he also published a lot of popular science books about marine organisms, such as jellyfish. He had theories about their development and evolution which he popularized in Germany. I make a connection between his work and the fantasies of the sea, between oil at the bottom of the sea and our consumption societies, whose dream material is plastic. In a way, plastic pursues the evolution in 'the forms of nature,' which as I said is now out of our hands. As once plastic is within nature, it's not in anybody's control, and it is metamorphosing in its own in a way that we can't observe, before coming back to human beings, in some way.

G.G: You suggest that in this process of evolution, plastic becomes an actor. Since we are evoking the relation between art and science, how do you position yourself towards the aesthetics of scientific research, the manner through which these aesthetics play a crucial part in revealing the world to us, but introduce also a bigger issue regarding the importance of 'representation' in the sciences?

T.N: I think visual artists and scientist are both trying to face the unknown, and that is a common point. But there is also a connecting point in the methods, particularly in the visual methods; many scientists use visual methods. Haeckel for example could draw very well, he was a very good draughtsman, and some of his theories are based on how he saw the world, for instance how embryos develop. He drew the images himself, which is a little unusual as a scientist at that time might have preferred to employ an artist to do the illustrations. But what I am saying is that it's more than illustrating, it's a mode of investigating reality, and there is a common thing in the visual arts and in science here. I am also thinking of Gaston Bachelard's *phenomenotechnique*, which is a kind of ontological method; it's not only a form of visualization, which to me is like illustrating something, it's how in a laboratory with a range of instruments, you make the phenomena materialize. The method is also a result, as the research and the result are interconnected in an ontological production, with often some visual trace of some kind. I am especially interested in how natural sciences integrate these traces.

G.G: Your work plays upon an understanding of science not as a mere reflection of reality, but as a construction of reality. In exploring traces, you have also made pieces that aim to represent a form of writing in nature, a writing by the wind (*Windtracers*, 2000) and by the waves (*Surf, 2001*). These explorations parallel an interest in early photography, photography as light writing, or *The Pencil of Nature*, to refer to the famous publication by Henry Fox Talbot.

T.N: Yes, in *Solarium* (2004) for instance I explore writing with light as a chemical process. To me photography is also a method of the natural sciences. That's why I mention the word traces too, because I am mostly working with analogue techniques, for practical reasons, and because I work outside. I mean I have nothing against digital techniques, it's a question of methods. As a lot of my work has to do with observation, and reporting, mostly outdoors, analogue photography provides a significant medium.

G.G: In terms of presentation, how do you decide on the mode of display? In *Frutti di Mare* for instance, you used photographs as well as rows of plastic containers in which the sculptures bathed in water, for *Mermaid's Tears*, you presented a necklace made of plastic pellets, and next to it you had installed plinths supporting collections of pellets and detritus, which were sorted and presented through a series of sieves.

T.N: I usually try to be a bit didactic. I want to give the spectator the possibility of understanding where the images come from. That's one of the reasons for showing the method and the apparatus that made the images. In the *Animal Cameras* (*Animalcams*, 2002), in which I explore the way in which different animals see the world, I showed the pinhole camera (the *sténopé*) I used to produce the images, together with the result, the photographic material that resulted from the dispositif. With this idea of *phenomenotechnique*, the aim is to show a whole cycle, from production to representation. Maybe the spectator sometimes might not understand the connection, but it doesn't matter, it's part of the work. And the methods are as important as the result, to me: the processes have to be shown. It's done in different ways, sometimes it's videos, sometimes it's the instrument, and sometimes all of them!

G.G: Your practice connects the activities of the artist and the scientist. In the past, these two activities and modes of thinking about the world, which are often seen as divergent, could be bound in revealing fashion. In the history of early maritime representation for instance, artists were instrumental in bringing to the fore the representation of nature, and could be linked to geography and scientific mapmaking. You have an interest for the representation of the natural world, and the emergence of facts. As such, the attention to processes leads to a reflection on the artifice, and 'artificial memories.' The leftovers made of plastic carry with them a narrative of interconnected history, that somehow bridges the gap between nature and the artifice: the plastic is the archetypal artificial material produced by men's industrial activities, but is ambiguously redeemed in its return to the sea, to the natural environment, where it pursues its own species' evolution.

T.N: I have been writing a bit about Catherine Malabou, who uses the term '*plasticité*' in the very general sense of the plasticity of the brain, and the same goes with plastic itself, it is malleable to a certain point, it depends on how the plastic objects are made, they can take any shape, but once it's done, the shape is more or less set, some plastic you can melt and reform, but some of them stay as such, and carry with them the shape that was given to them. I experienced this aspect first hand, as I had to mechanically put these plastics together. I also tried to melt them with an iron, but I often failed as they don't really stick together, and I had to use mechanical connections. Something similar applies metaphorically, in relation to evolution and how the material behaves.

With reference to the optical instruments, the notion of artifice is very negative. The artifice is something that the instrument produces, a biproduct that is not real. And to some extent any research product contains an artificial character. And the whole instrument can be understood as an artifice. For example in the work I did on the River Senne in Brussels, I used a very poor kind of toy microscope. What it produced was a lot of artifice, because of the chromatic aberration of the lens. The objects I was looking at were surrounded by colours that did not belong to the object, they were an artifice of the lenses, and I recorded those artifices very meticulously in my watercolour paintings. I was as much interested in the artifice that the instrument produced, as I was in the real optical image, thereby investigating the role of technology in transforming reality.

G.G: The idea of 'the artifice' plays a mediating role, as an image or a metaphor, rather than an essence of artificiality to be opposed to the 'natural.' The categories of 'artifice' and 'nature' are explored through the way they are constructed, created.

T.N: Haeckel somehow understood this, *Kunstformen der Natur* was very influential in the field of design, and architecture of the time. It's a romantic idea, the creating nature.

G.G: In the Amos Anderson Art museum, I saw a tapestry by Greta Skogster-Lehtinen (1900–1994) depicting the harbour of Helsinki, a very large and modern design from the 1930s. The natural landscape depicted in this tapestry within which Helsinki dwells, is, however, being colonized by urban extensions. Similarly, you have explored the progress of asphalts in an early work, *Ulysses in Helsinki* (1999), in which you mapped the extension of Helsinki towards the sea, that brought islands to become attached to the mainland. You are pursuing this strand in combination with your interest in plastic through an attention to the city's current underwater expansion, through the recuperation of underbelly explosives.

T.N: Presently the cities of Helsinki and Espoo are building a new underground line, which requires exploding the rock, then they ship the rest into the sea. They use this explosive plastic substance, in bits and pieces between the gravel. It's been much discussed in the press, as it's a new form of plastic pollution. I'm using the anti-slip plastic mats as a support structure. I use a crochet technique. There is a very specific Finnish tapestry tradition, called, 'ryijy,' or Finnish rugs that were woven by peasant women in the 19th century. The rugs were used on the walls for isolation, as well as decoration. The tapestry you referred to evokes this anonymous tradition of Finnish rugs, as it was revisited by textile artists, women as well, and yes you could say I am also revisiting the tradition! I'm not using only bags, but I also collect pieces of ropes, and pieces of plastic shock tube detonator that is used in exploding the rock. I'm planning to weave and to crochet these bags to make a romantic seascape, in a Turner style, sunset on the sea!

Note

1 This conversation took place in Helsinki in August 2016. www.tuulanarhinen.net

8 Naturally Representative

The Environmental Planning of the New African Capitals Abuja and Dodoma

Ayala Levin

In the 1970s, North American architects designed two new capital cities in postcolonial Africa: Dodoma in socialist Tanzania by the Canadian firm Project Planning Associates and New York partners William Conklin and James Rossant, and Abuja in westernized Nigeria by the Philadelphia firm Wallace, McHarg, Roberts and Todd. The period, dubbed in American history as the "Environmental decade," is characterized by an increased attention to the limitations of energy and food resources in the global north and the global south respectively. This "environmental turn" had a dual effect on North American firms of architecture and planning: at the home front, the federal government increased environmental regulations, and concurrently, the energy crisis drove them to seek projects abroad, particularly in oil rich countries. In cases where these projects were funded by the US government, these too had to be subjected to the federal government's environmental regulations.[1] Working abroad in unfamiliar territories and extreme climatic conditions did not only present North American architects with opportunities to expand their practice during an economic stagnation but also to experiment with a range of extreme environmental conditions.

This environmental turn coincided with a shift that took place in the 1970s in the Western donor countries' approach to development aid, spurred by the world food crisis of 1972–1974. In 1973 former US Secretary of Defense and President of the World Bank Robert McNamara addressed the Bank's Board Governors with a pledge to devoting its resources to small-scale agricultural production. Other donor countries followed suit to divert their aid from industrial to agricultural development. The World Food Conference of 1974 cemented this shift to food security.[2] The governments of Tanzania and Nigeria were diametrically opposed in their reactions to these shifts in development thinking—a combination of their natural resources and ideological leaning. For Nigeria, who experienced in the 1970s an unprecedented wealth as an oil rich country, the petrodollars euphoria turned the political elite's attention away from agriculture, and the hardships of those who did not enjoy the fruits of the country's riches. In contrast, Tanzania's charismatic leader Julius Nyerere embraced this shift of emphasis on agriculture, as it was aligned with the vision he had been developing since the 1960s for an African socialist rural-based society. An agriculture-based path to an African modernity, he believed, was a more sustainable alternative to the promises of industrialized modernization.

The planning of the new capitals in the countries' hinterland presented opportunities to control the new cities' growth, and preempt the various ills associated with the previous capitals and colonial port cities, Lagos in Nigeria and Dar es Salaam in

Tanzania. The relocation to the hinterland, where sparsely populated large tracts of land were available, was to reverse the history of colonial exploitation that concentrated wealth in the port cities, as a connecting node in an economy based on extraction. While these concerns were shared widely in postcolonial planning and had been experimented with since the late 1950s with the planning of the capitals of Brasilia and Chandigarh, to note the most prominent examples, planning in the 1970s was qualitatively different since the lessons of these attempts have been critically studied and incorporated into the plans. The opportunity to build from scratch in Tanzania's and Nigeria's hinterland, in other words, was not simply another "high modernist" iteration of abstract planning,[3] but a response to the concrete results of contemporaneous projects, and the growing (Western) concerns over the environment. While the two projects firmly belong in the post-war trajectory of foreign development planning which was based on vast territorial transformations, they also represent a turn to a more moderate and cautious approach to modernization's faith in the positive transformative capacity of natural environments by rational planning and technology. This shift is expressed by the involvement of Ian McHarg of WMRT, the prominent ecological planner, in the drafting of the master plan for Abuja. In the Dodoma case, this shift is expressed in the response of the planners, the Toronto firm Project Planning Associates and the New York partners William Conklin and James Rossant, to Nyerere's administration's demands for a capital city that would reflect Tanzania's ideology of self-reliance.

The representation of nature as resource for consumption or production is essential to the ordering of society and geography—the objects of development planning.[4] This chapter asks how concerns over the ordering and management of resources affected the design of the capitals and are reflected in their representational programs. In other words, this chapter analyzes how the planners incorporated the country's natural environment in their visions for the new capitals vis-à-vis the governments these capitals were meant to represent. More specifically, this essay is concerned with the role of the representation of nature in the city in the formation of postcolonial citizenry, and in the promotion of the countries' social and economic development agendas to international audiences as varied as potential investors, tourists and development agencies. This emphasis on the new capitals' relationships to their natural environments and resources sets this essay apart from previous accounts that focused on the representation of power at their monumental cores, particularly the relationship between the capitol hill complex and the city.[5] While Abuja's and Dodoma's planning certainly owed to this modern tradition of representation of political power, they also differ in their reference to American new towns and their suburban landscapes, of which their respective planners had direct knowledge and experience.[6] The two traditions, the planning of capital cities and New Towns, are incompatible in terms of both scale and purpose, as one traditionally emphasizes civic monumentality, based on a hierarchical ordering and radial axes, while the other favors the picturesque blending of low-rise housing and non-monumental community amenities. Due to their distinct emphases, these approaches differ in their incorporation of the natural environment, or its signifiers, in their planning objectives. This essay explores how the architects attempted to reconcile these distinct and incompatible models in the capitals' integration to the site, on the one hand, and in attributing it a national representational character, on the other.

Relocation

In 1975, just one day after the coup that brought him to power, General Murtala Muhammad announced the relocation of the Nigerian capital from Lagos. Suffering since the mid-1960s from a succession of coups and a bloody civil war, the governments that succeeded Muhammad's short-lived rule carried out his vision to construct the new capital, which they deemed a panacea for the divided country (from three federated states in 1960, the year Nigeria received its independence, it fragmented to 19 states in 1976). The successive governments explained that Nigeria was in a process of building "a nation out of many nations." The capital city was to provide a fresh start; a new cultural base for a national unity in a supposedly virgin land, undisturbed by ethnic rivalry.[7]

To convey impartiality, the government assigned a committee to examine the need to relocate from Lagos, and to suggest alternative sites. The committee listed several reasons for the relocation from Lagos: first, its peripheral location in relation to the country and its identification with the Yoruba ethnic group who dominated the region; second, the relocation was to relieve it from its dual role as the capital of the federal government and Lagos state; and finally, its congestion. As a major port, an industrial, banking and commercial center built on lagoons and creeks, Lagos suffered major traffic problems.[8] In addition, this congestion was deemed a threat to the security of the governing elite.[9]

The committee considered 33 sites following a list of criteria, whose top four were: centrality, health and climate, land availability and use, and water supply. The 8,000 square kilometers—more than twice the territory of Lagos state—chosen for the site are in the Middle Belt region, north of the confluence of rivers Niger and Benue, south of the town that used to be named Abuja.[10] Its predominantly savanna area was not as agriculturally productive as the forest regions north and south of it. Known to be infested with tsetse flies, the area was neglected for generations. With no strong social organization, its ethnic composition was extremely fragmented. Although it consisted of less than 20 % of the country's population, it accounted for over 70 % of its ethnic groups. Ironically, it is this economic and social under-development that proved to be an advantage in its selection as a site. Not associated with any specific ethnic group, it offered a neutral ground—a buffer zone between the ethnically and religiously divided north and south of the country.[11] Like in Brasilia, the building of the new capital in the middle of the country was presented as part of a large infrastructural development that would better connect the country politically and economically. In addition, its location inland far from conflict-ridden regions was supposed to assist the government in handling civil conflicts.

In the summer of 1977, an international advisory committee was formed to work with the Nigerian committee to design a city earmarked for 3.1 million. The Philadelphia-based firm WMRT, who has had experience in developing a comprehensive landscape plan for Washington the previous decade, was chosen to determine the site and urban form. The firm approached the task by conducting a comprehensive "environmental inventory," foregrounding climate, geology, physiography, hydrology, soils, vegetation, wildlife and disease, land use, and agriculture in the selection of the site within the Federal Capital Territory.[12]

The relocation of the capital in Tanzania was similarly motivated by a growing congestion in Dar es Salaam, but it was painted with a distinct anti-urban rhetoric.

Like the "slum clearance" and urban planning attempts in 1960s Lagos, the Tanzanian government considered the re-planning of Dar es Salaam prior to the decision to relocate the capital.[13] Project Planning Associates, the same firm hired for the Dodoma project, had been hired in 1965 for this purpose. However, in 1967 president Julius Nyerere laid down his policy for African socialism and self-reliance known as the Arusha Declaration. Though partly inspired by China, this policy promoted rural community life based on African principles of *ujamaa* (translated as familyhood in Swahili). In this vision, the African village was reimagined as part of an economically and culturally independent rural global south,[14] and presented a radical alternative to the precepts of industrial-based modernization that both Cold War superpowers prescribed to. With the overwhelming majority of Tanzania's population living in rural areas, the country's economy would continue to be based on agriculture, while industry was to be introduced at small scale, in support of agricultural production. As an important harbor that connected East Africa with the Indian Ocean, Dar es Salaam was distinctly cosmopolitan and did not fit the national *ujamaa* ideology. In addition to being associated with the corrupt morals of the city, which was now condemned as inauthentic to African way of life, its port carried a history of colonial exploitation.[15]

With its emphasis on rural living, the Arusha declaration marked a watershed in urban planning in the country. Following the declaration, the planning of Dar es Salaam was sidelined in favor of a national villagization campaign. The country was imagined as a nation of ujamaa villages, where the people could cooperate in small groups, and these, in turn, would cooperate as joint enterprises coordinated by the state. In 1972, the government decided to relocate its seat to Dodoma to crown the extensive villagization campaign it conducted across the country, including in the draught-stricken Dodoma region, as "the chief village among a nation of villages."[16]

As in Abuja, the choice of site was determined by a government committee for political and practical reasons. the existing town of Dodoma was a stronghold of Tanganyika African National Union (TANU), the ruling party. Established in 1910 following the construction of Tanganyika Central Line, the town is located on one of East Africa's main caravan routes from the interior to the coast of the Indian Ocean, at the crossroad of six roads.[17] Its central position and political affiliation made it a perfect candidate to serve as the headquarters from which the government could project its control and extend its services directly to the rural areas. Designated for 350,000 (with the capacity to grow to 750,000) inhabitants, Dodoma's plan was far less ambitious in scope than Abuja's projected 3.1 million. If the challenge for the Abuja planners was to devise a unifying image, while allowing for an incremental growth, the challenge in the planning of Dodoma was to create a socialist city that would communicate rural ideals on the one hand, and exhibit in built form its ideology of self-reliance.

Symbolic Core

The planners of Abuja synthesized their "ecological inventory" to determine urban form. The dictates were "man made constrains," mainly the future site of the international airport, and the "dominant visual elements." While the first could be determined by empirical calculations, such as wind and land suitability, decisions on the visual qualities of the site left room for cultural predilections and norms. Unlike in eastern cultures, such as the firm met in Tehran where they designed the Pardisan

Environmental Park in 1973–1975, in Nigeria landscaping was a foreign concept, which had a limited effect on Nigerian towns via British colonial planning.[18] Following the previous examples of Washington, Canberra, Chandigarh and Brasilia, the siting and urban form was to correspond and capitalize on the natural features, even, as in the case of Brasilia's lake, or Chandigarh's capitol hill, these were artificially created. In Abuja, the imposing Aso Hill, a granite rock, was chosen as the dramatic backdrop for the federal government triangle, the Three Arms Zone (Figure 8.1). The latter, separated by a valley, crowned an informally landscaped mall, which was flanked by ministries and other governmental, commercial and cultural institutions, and terminated with a transportation hub. While this ordering clearly draws from Brasilia's Plano Piloto, it differs in the use of the natural topography in the siting of its main buildings. These were sited on a series of hill crests that gave accent to the axis. The dramatic effect of the axis was therefore achieved via the careful placement of the buildings and open spaces in relation to the topography of hills and valleys, serving either as platform (the series of hill crests), separation zones (the valley facing the Three Arms Zone) or as national monument (Aso Hill).

As in their study for a landscape plan for Washington DC, the WMRT planners considered the geological features of the site as indicative of a *longue durèe* history that preceded human intervention. Perceiving the genius loci (a term they used for Washington but not for Abuja) in terms of natural formations allowed the planners to use ostensibly neutral forms, the result of "million years of geological history," ostensibly unaffected by ethnic rivalries such as the ones surrounding the nearby Zuma rock.[19] At the same time, the natural features of the site were not so dominant as to "impose a constraint on straight streets nor produce a sense of natural order." The planners did not seek to recreate the latter. By superimposing a grid system on the rolling terrain, they sought to produce "a unique juxtaposition of qualities in which disorder has been overcome."[20] This control would reach its apex in the botanical garden-arboretum, which the planners located between the Three Arms Zone and Aso Hill "where specimen trees and flora from Nigeria and West Africa would be arranged

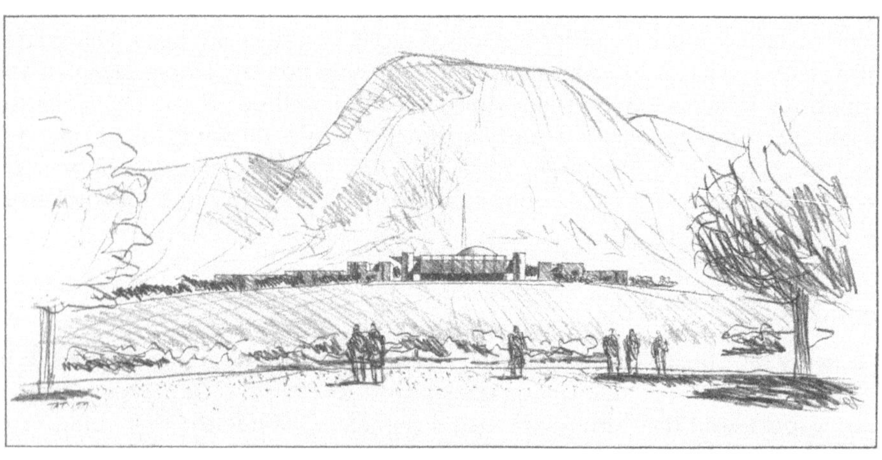

Figure 8.1 Thomas A. Todd, Aso Hill Framing the Three Arms Zone, Abuja, 1979. Ian L. McHarg Papers, The Architectural Archives, University of Pennsylvania.

in a controlled naturalistic setting as the Gardens of the Parliament Buildings."[21] Outside of the city, in its outskirts, nature could reign completely and be protected as national monuments. Seeing this as an opportunity to increase national revenue from international tourism, a game reserve was to be built in the Bobo Plains south of the city, along the lines of the safari in the outskirts of Nairobi. Serving the dual and somewhat contradictory interests of conservation and sport hunting, it was suggested that the park reservations would even function as a "genetic bank to provide a living source that will be of value at some later date."[22]

Planning Projects Associates followed Brasilia's Plano Pilato even more closely than their counterparts in Abuja in their plan for Dar es Salaam a decade earlier, where they used a strong bird-like image to orient the city toward the port. In Dodoma, however, they could no longer repeat this hierarchical scheme as their task was to plan "the first nonmonumental capital city."[23] With no central image or axis to define the plan, the planners' attention focused on the residential communities. With the existing city functioning like a nucleus, a network of residential units connected by curvilinear roads joined it to create a multi-focal plan. In effect, the planners relegated the governmental center to the periphery of the plan, where it appears almost as an afterthought (Figure 8.2).

Designed by William Conklin and James Rossant in 1979–1980, Dodoma's new representative mall is located as a midway center between the existing city, the new residential network, and the parliament and presidential palace located on a hill further east.[24] Characterized by mixed-use and dense construction, where government ministries are interspersed with a hotel, restaurants, shops, housing and cultural institutions, the mall is designed to combine the "excitement of the village and the intensity of communication of the big city."[25] To provide for shade and facilitate pedestrian movement, the buildings are porticoed and the open spaces are heavily planted. Among the building regulations Project Planning Associates specified were the use of local materials—brick, tile, concrete and local timbers—as well as local techniques. "Particular emphasis shall be placed on the imaginative use of Tanzanian materials and skills," the planners stated, while warning that "exotic imported techniques should be avoided."[26] This ideology of self-reliance is further expressed in the limitation of building height to no more than three stories, and the orientation of buildings to maximize ventilation and daylight. These regulations aimed to eliminate the need for elevators and air conditioners, and to keep electrical consumption to the minimum.[27] Considering these regulations, the planners' emphatic rejection of the skyscraper in favor of low horizontal buildings was not only a symbolic gesture, as others have interpreted it, but pertained to the country's limited resources.[28]

The ideology of self-reliance elevated the country's ability to cope with its meager natural resources into a value, which Conklin and Rossant shrewdly employed as a representational strategy. This strategy is expressed in the hierarchical disposition of the ministries that symbolizes their relative position in the narrative of Tanzania's progress, from past to future. The civic procession starts from Natural Resources and Agriculture ministries on the first terrace, continues via Foreign Affairs and Finance on the second and third terraces, to Lands, Housing and Urban Development, and Water and Energy on the fourth terrace. The schematic plan indicates ascendance from the ministries of resources (terraces 1–2), through the ministries of infrastructure (terraces 3–4), to social ministries (terraces 5–6), while the policy ministries are crossing at the center and mixing with the others "in a symbolic knit of unity."[29] The narrative of

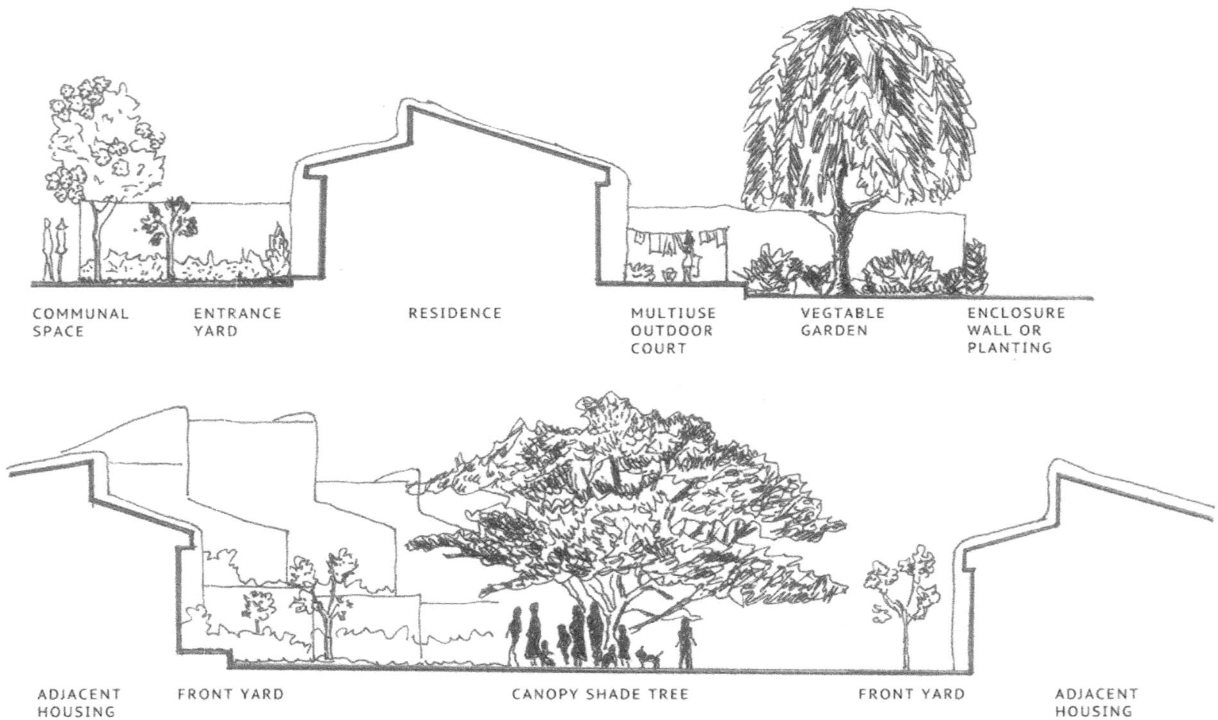

Figure 8.2 Planning Projects Associates, Dodoma, Master Plan and Section of Residential Yards, 1977, hand-drawn reproduction following an image in "Dodoma Capital City," *Landscape Architecture* (July 1977): 327. Macklin Hancock/Project Planning Associates Ltd. fonds, Archival & Special Collections, University of Guelph Library.

ascendance from agrarian economy via water, energy and housing infrastructure, and entry into world economy via international affairs and banking does not stipulate that one stage overrides the other. Rephrasing the Marxist critique of the inversion of the material basis and the ideological superstructure, Nyerere polemically posed in 1968 the questions: "what is the foundation" and "what is the fruit of development." According to him, "between MONEY and PEOPLE it is obvious that people and their HARD WORK are the foundation of development, and money is one of the fruits of that hard work." Rather than promising the miracle of an accelerated growth that would help Tanzania "catching up" with the developed countries, as postulated by industrial modernization, Nyerere explained that "Industries will come and money will come but their foundation is THE PEOPLE and their HARD WORK, especially in AGRICULTURE. This is the meaning of self-reliance."[30] Conklin's and Rossant's mall embodied the arduous path that the Tanzanian citizens are facing, with each of the terraces is built on the base of the previous one to preclude the capitalist inversion which Marx had characterized as a camera obscura, and which Nyerere updated to characterize the false ideology of modernization: "From now on we shall stand upright and walk forward on our feet rather than look at this problem upside down."[31] The phased construction of each terrace as a "complete, finished statement"[32]—so that the mall's incompletion would not be interpreted as failure—further reinforced this massage. Giving physical form to the policy of self-reliance, the terraced mall completed the political work of curbing the citizens' expectations for economic development while harnessing and mobilizing them as agricultural workforce.

The cumulative experience of the pedestrian procession starts at the Freedom Square at the northern edge of the mall, and reaches its apex at the southern end of the mall with the People's or Ujaama Square, crowned by an abstracted sculpture of an acacia tree. Drawing on the symbolic function of the acacia tree, under its shade Tanzanian villagers habitually congregate, the sculpture marks the space where the nation can congregate and enact its traditionally-based communality. As a sculptured abstraction, the tree symbolizes the convergence of the natural and the social, or the imagined authenticity of rural life with the bureaucratic structure of the state.

This representational strategy of conflating the natural with the social did political work also in relation to the international guests, upon whose financial and technical aid Tanzania depended to realize the natural potentials of the country.[33] A sunken tropical garden, bridged to provide a dramatic entrance to the Agriculture Ministry, represented the vegetation of Tanzania's coastal area, while a restaurant served the dishes associated with the area. An additional garden representing the fertile northern region led to the Agriculture Ministry. The recreations of these remote environments, as well as the decorative pools that accompanied them, relied on a high volume of water supply that stood in conflict with the ideology of self-reliance.[34] Their inclusion in the plan despite their high maintenance testifies to the importance in conveying the country's potential riches to the international in addition to the local communities. For the same purpose, Iseni Park, a large park west to the mall in the area adjacent to the embassies, was designed with carefully composed vistas that introduced variety to the otherwise bland savanna landscape.

Residential Districts

In Abuja, drawing from Brasilia, the residential sectors extend from both sides of the monumental axis. Learning from the lessons of Brasilia, however, the planners of Abuja attempted to preempt the mushrooming of squatter settlements in the outskirts of the city by planning for much larger residential sectors comprising between 100,000–250,000 inhabitants, which could accommodate various incomes, as well as Nigerian lifestyle, whose household was predominately based on the extended family.[35] In contrast to the grid system at the central monumental spine, the residential sectors are defined by the picturesqueness of curved or non-geometric street patterns. These sectors are subdivided into districts of 30,000–60,000, which again are subdivided into neighborhoods averaging 3,000–5,000 residents each, to accommodate a "strong sense of local residential community" such as the planners found in existing Nigerian towns.[36]

Connected by a parkway, the sectors are a scaled-up version of the New Town Woodland, TX, which McHarg planned in the early 1970s for a community of 180,000.[37] In the Nigerian context, the American parkway that provides scenic vistas to the car drivers was extended to provide outdoor spaces for schools, burial grounds and open spaces for the gatherings of lorries and informal markets. In addition, similarly to Woodland which was planned as a New Town in the periphery of Houston, the sectors did not function as dormitory suburbs. In addition to the typical residential amenities found in neighborhood units such as schools and shops, the sectors provided employment centers, ranging from research centers to small-scale industries. Green spaces within the sectors were available for recreation, pedestrian or bicycle circulation, and some plots were to be allotted for community vegetable gardens.[38] However, the management of the sectors was given to Milton Keynes Management Corporation, who emphasized car circulation, shopping centers and Westernized housing that could neither accommodate the extended family households, nor be available for purchase by any other than the Nigerian higher middle class.[39]

In Dodoma, the residential units were not secondary to the political and commercial center. Each of the units consisted of four clusters of approximately 7,000 each, joint together by a community center and surrounded by a continuous park-belt. The planners rejected gridiron as well as radial plans. Instead, the streets layout responded to the topography to minimize intervention in the landscape, both in terms of its physiography and its agricultural capacities. The built areas were located on the land least suitable for agriculture or water storage, to maximize arable land in the surrounding area.[40] The journal *Landscape Architecture*, which awarded the project a merit award in July 1977, explained that to maintain a rural character, the planners emphasized the residents' close connection to the land: "The landscape heritage of Tanzania pervades consistently through all planning. At the village [residential] level, great consideration is given to urban agriculture; at the civic level indigenous trees, shrubs and grasses are brought into the urban open space, road systems, parks and plazas."[41] The open space of the landscape was supposed to flow through the city, so that every community would be close to major parkland, hills or green space.

The relationship between family, land and community is intimately expressed at the scale of the "TANU ten-unit cell," the smallest organizing unit of the residential districts. It consisted of ten adjacent detached houses (each for a family of approximately five) grouped around a shared common space. Like in the ujaama village, where the communal space is a disproportionally large space compared to the individual plots, the common space was the most important aspect of the TANU ten-unit cell.[42] The feature that represents this communal space, as in the Ujaama Square, is the acacia tree under its canopy the residents congregate (Figure 8.2).[43] While the differentiation between the front and back yards, the first public and the latter private, situates this housing as a transitional space for the cultivation of the nuclear family following Western suburban pattern, it nonetheless has some connections to some patterns in Tanzanian traditional villages, where homesteads are enclosed and separated from each by a clear boundary.[44]

The TANU cell's common space differentiates Dodoma's residential districts from the suburban planning of Abuja, where following their North American and British examples, the communal spaces were supplanted by the shopping centers. In effect, neighborly encounters or children's play is relegated to the interior road and parking spaces (Figure 8.3). If in the planning of Abuja, the planners needed to quickly correct their early stipulation that only 30% of the residents will own cars, then in Dodoma, where an innovative bus and bicycle lanes' system was put in place in accordance with the ideology of self-reliance, mass ownership of cars was considered neither possible nor desirable.[45] Despite this empathic rejection of car ownership, the planners designated a parking space in the TANU cell plan, which for the most part left unpaved. This curious decision was a deliberate attempt to preempt the possibility that the parking area would take over the communal space in the distant future.

Conclusion

In designing the national capitals, the planners considered both domestic and international audiences. For the international guests, the natural scenery was to represent images of abundance and untapped riches. In Dodoma, these were presented as resources waiting to be cultivated. In Abuja, the natural environment was exoticized to offer wild natural reserves that could cater competing desires such as game hunting, on alternatively, the preservation of a genetic repository in preparation for a looming global ecological disaster. In Africa's small-scale societies where natural resources are consumed for livelihood, or associated with the spiritual world, the association of nature with aesthetic or recreational consumption is luxury at best, and desecration, at worst.[46] The creation of public parks, gardens and vistas served to educate the taste of domestic audiences, and cultivate civic pride in their countries' natural environment. However, the relationship between the representation of the environment and

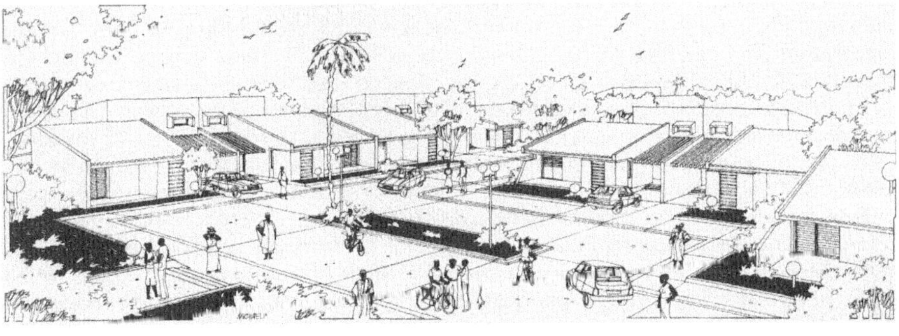

Figure 8.3 Egbor Arenco Planning and Yasky Architects, a Cluster of Row Houses for Medium Density Areas, IJKL, Abuja, 1982

Courtesy of Amos Spitz

the character of the public spaces differed greatly between Abuja and Dodoma, and prescribed different forms of civility in relation to the different paths of modernization the countries subscribed to.

The carefully composed vistas of the Iseni Park and the landscaped terraced mall in Dodoma, similarly to the arboretum and the gridded axis in Abuja, served to convey to domestic audiences their governments' prowess in managing their natural resources. As part of this ordering, the local audiences would learn to cultivate an aesthetic appreciation of their local environment by school visits to national parks, and driving through such monuments like Zuma Rock in the case of domestic tourism in Abuja, or, for its residents, by the daily parkway commute to the government center. In Dodoma, the planners imagined how the residents would enjoy family picnics in Iseni Park, or the local landscape from vistas in their neighborhoods and the stepped mall. In both cases, the natural environment served as a unifying force for multi-ethnic populations. However, while in Abuja it was raised to the level of national monuments to bypass the country's extreme social unrest and inequality, in Dodoma, the monumentalizing of the acacia tree did not attenuate its social value, but capitalized on it to metaphorically link between the state and its population. The difference in the character of the "social" is manifested in the degree of access to nature in each of the cases. While in Abuja it is based primarily on the car, and therefore is economically stratified, in Dodoma the access is primarily pedestrian, and thus more egalitarian. In Abuja the relationship to the environment is based on the values of consumerist society the plan promotes, based on the country's abundant oil. In stark contrast, the relationship between people and the land in Dodoma emphasizes its inherent reciprocity as the social meaning of the environment derives from its economic capacity, and vice versa, the economic capacity of the environment is determined by the social.

Notes

1 Jeffery W. Cody, *Exporting American Architecture, 1870–2000* (London; New York: Routledge, 2003), 153.
2 Christopher D. Gerrard, *Promoting Third World Agriculture: Lessons of Recent Experience* (Ottawa: North-South Institute, 1983), 1–2.
3 As James C. Scott famously argued. See James C. Scott, *Seeing Like a State* (New Haven, CT; London: Yale University Press, 1998), 103–146.
4 Ibid; Timothy Mitchell, "The Object of Development: America's Egypt," in *Power of Development*, ed. Jonathan Crush (London; New York: Routledge, 1995), 129–157; Christopher Bonneuil, "Development as Experiment: Science and State Building in Late Colonial and Postcolonial Africa, 1930–1970," *Orisis* 1, Nature and Empire: Science and the Colonial Enterprise (2000): 258–281. Development projects' reliance on foreign technical expertise and funding has led to criticize them as foreign implantations, whose grand scale and expenditure were meant to cater the new Westernized tastes and habits of the local elites at the cost of perpetuating inequality and destroying the ecology and vernacular modes of livelihood. This paper introduces a more nuanced reading of the interaction between local governments and foreign experts in the construction of their urban environments.
5 Lawrence Vale, *Architecture, Power, and National Identity* (New York; London: Routledge, 2008), 146–193; For an extensive and critical account of the planning of Abuja see Nnamdi Elleh, *Abuja: The Single Most Ambitious Urban Design Project of the 20th Century* (Weimar: VDG, 2001).
6 The planners of Dodoma and Abuja were involved in the planning of some of the most important New Towns and suburban developments in the 1950s–1970s. Ian McHarg planned Woodlands, TX. Conklin and Rossant planned Reston, VA. Macklin Hancock, the head of Project Planning Associates, designed Toronto's Don Mills suburban community.

7 Abuja, *The Nation's New Capital* (Lagos: Department of Information, Office of the President, 1975).
8 Federal Capital Development Authority, *Abuja: The New Federal Capital of Nigeria* (Abidjan, Ivory Coast: Société interafricaine d'édition, 1983), 18–19.
9 Elleh interprets Abuja as a flight of the Nigerian elite from the political volatility of Lagos. Elleh, 77–78.
10 The original town of Abuja, founded in 1828 by Emir Abu Ja was demolished and rebuilt elsewhere under the new name Seluja to avoid confusion with the capital.
11 Although this was the official reasoning, Abuja is closer to the northern Islamic power base of the military regimes of the 1970s, 1980s and 1990s.
12 On McHarg's "environmental inventory" see Anne Whiston Spirn, "Ian McHarg, Landscape Architecture, and Environmentalism: Ideas and Methods in Context," in *Environmentalism in Landscape Architecture*, ed. Michel Conan (Washington, DC: Dumbarton Oaks Research Library and Collection, 2000), 107–108.
13 For "slum clearance," in Lagos see Daniel Immerwahr, "The Politics of Architecture and Urbanism in Postcolonial Lagos, 1960–1986," *Journal of African Cultural Studies* 19, no. 2 (December 2007), 170–179. For an urban plan proposed by UN consultants see: UN Technical Assistance Administration, *Metropolitan Lagos: Prepared for the Government of Nigeria* (New York: United Nations, Commissioner for Technical Assistance, Dept. of Economic and Social Affairs, 1964).
14 Emily Callaci, " 'Chief Village in a Nation of Villages': History, Race and Authority in Tanzania's Dodoma Plan," *Urban History* 43, no. 1 (2016), 104–105.
15 Ibid., 96–116.
16 Callaci, 97–98. On the villagization campaign see Scott, 223–261; John Shao, "The Villagization Program and the Disruption of the Ecological Balance in Tanzania," *Canadian Journal of African Studies* 20, no. 2 (1986), 219–239.
17 Capital Development Authority of Tanzania. *A Portrait of Dodoma* (Dodoma: The Authority, 1975), 31–45.
18 Kathleen John-Alder, "Paradise Reconsidered: The Early Design History of Pradisan Park in Tehran," in *Contemporary Urban Landscapes of the Middle East*, ed. Mohammad Gharipour (London: Routledge, 2016), 120–148; Liora Bigon, "Garden Cities in Colonial Africa: A Note on Historiography," *Planning Perspectives* 28, no. 3 (2013), 480–481.
19 Nigeria, The Federal Capital Development Authority, *The Master Plan for Abuja, the New Federal Capital of Nigeria* (Lagos: The Federal Capital Development Authority, 1979), 85. The Zuma rock is associated with the history of the Zuba people, and perhaps others. See http://zuba.org.ng/historyofzumarock.html.
20 Ibid., 83.
21 Ibid., 86.
22 Ibid., 256–257.
23 Vale, 179.
24 Vale, 186–188.
25 C.G. Kahama, William Conklin and James Rossant, *Urban Design for the National Capital Centre* (Dodoma: Capital Development Authority of Tanzania, 1980), 13.
26 Project Planning Associates, Site Planning Building Guide, XL3 MS A021020, Macklin Hancock/Project Planning Associates Ltd. fonds, Archival & Special Collections, University of Guelph Library.
27 Ibid; Kahama, Conklin and Rossant, *Urban Design for the National Capital Centre*, 6.
28 For previous interpretations see Luce Beeckmans, "Continuities with the Past and Future: The Architecture of Nation-Building in Kinshasa (DR Congo) and Dodoma (Tanzania) in the Post-Independence years 1960–1970" (paper presented at the Crossing Boundaries: Rethinking European Architecture Beyond Europe Conference, Palermo, April 13–17, 2014); Duncan Yoon, "The Rationalization of Space and Time: Dodoma and Socialist Modernity," *Ufahamu: A Journal of African Studies* 36, no. 2 (2011), 1–27.
29 Kahama, Conklin and Rossant, *Urban Design for the National Capital Centre*, 17.
30 Cited in Yoon, 8.
31 Ibid.
32 Kahama, Conklin and Rossant, *Urban Design for the National Capital Centre*, 10.

33 As other development projects, the building of Dodoma was anything but autonomous or "authentically local" as it was supported by donors and technical assistance from Denmark, India, Australia, West Germany, Mexico, Italy, Indonesia and Austria, in addition to the planners from Canada, the US and, later, China. See Callaci, 111. For an analysis of the paradox of depending on foreign aid to realize Tanzania's ideology of self-reliance see Linda Freeman, "CIDA, Wheat, and Rural Development in Tanzania," *Canadian Journal of African Studies* 16, no. 3 (1982), 479–504.

34 Considerable funds were spent on a sewage system nicknamed "Rolls Royce Sewers." Luce Beeckmans, "The Architecture of Nation-Building in Africa as a Development Aid Project: Designing the Capital Cities of Kinshasa (Congo) and Dodoma (Tanzania) in the Post-Independence Years," *Progress in Planning* (2017): 8.

35 The Master Plan for Abuja, 69–70.

36 Ibid., 69.

37 On the planning of Woodlands see Anne Forsyth, *Reforming Suburbia: The Planned Communities of Irvine, Columbia, and the Woodlands* (Berkeley: University of California Press, 2005), 161–207.

38 *The Master Plan for Abuja*, 128.

39 Elleh, 75–76.

40 Ibid., 180.

41 "Dodoma Capital City," *Landscape Architecture* (July 1977), 326.

42 Shao, 233; For an image of a plan see Scott, 243.

43 Like the ujaama village, the TANU cell was to be managed as a cooperative by a cell leader, who would serve as relay between the residents, the party and the state. For a critique of this supposed "grassroots participation" see Scott, 237.

44 Shao, 228.

45 The original plan envisioned a 70% reliance on public transport versus 30% reliance on private cars but this was quickly changed to accommodate the residents' reliance on private cars.

46 For an example of the relationship between the social, the spiritual and the natural world in a small-scale African society see Zoe Strother, "Architecture against the State: The Virtues of Impermanence in the Kibulu of Eastern Pende Chiefs in Central Africa," *Journal of the Society of Architectural Historians*, 63, no. 3 (September 2004), 277–278.

References

Abuja. *The Nation's New Capital*. Lagos: Department of Information, Office of the President, 1975.

Beeckmans, Luce. "Continuities With the Past and Future: The Architecture of Nation-Building in Kinshasa (DR Congo) and Dodoma (Tanzania) in the Post-Independence Years 1960–1970." Paper presented at the Crossing Boundaries: Rethinking European Architecture Beyond Europe Conference. Palermo, April 13–17, 2014.

Beeckmans, Luce. "The Architecture of Nation-Building in Africa as a Development Aid Project: Designing the Capital Cities of Kinshasa (Congo) and Dodoma (Tanzania) in the Post-Independence Years." *Progress in Planning* (2017): 1–28.

Bigon, Liora. "Garden Cities in Colonial Africa: A Note on Historiography." *Planning Perspectives* 28, no. 3 (2013): 477–485.

Bonneuil, Christopher. "Development as Experiment: Science and State Building in Late Colonial and Postcolonial Africa, 1930–1970." *Orisis Nature and Empire: Science and the Colonial Enterprise* (2000): 258–281.

Callaci, Emily. " 'Chief Village in a Nation of Villages': History, Race and Authority in Tanzania's Dodoma Plan." *Urban History* 43, no. 1 (2016): 96–116.

Capital Development Authority of Tanzania. *A Portrait of Dodoma*. Dodoma: The Authority, 1975.

Cody, Jeffery W. *Exporting American Architecture, 1870–2000*. London, New York: Routledge, 2003.

"Dodoma Capital City." *Landscape Architecture* (July 1977): 326–327.

Elleh, Nnamdi. *Abuja: The Single Most Ambitious Urban Design Project of the 20th Century*. Weimar: VDG, 2001.

Federal Capital Development Authority. *Abuja: The New Federal Capital of Nigeria*. Abidjan, Ivory Coast: Société interafricaine d'édition, 1983.

Forsyth, Anne. *Reforming Suburbia: The Planned Communities of Irvine, Columbia, and the Woodlands*. Berkeley: University of California Press, 2005.

Freeman, Linda. "CIDA, Wheat, and Rural Development in Tanzania." *Canadian Journal of African Studies* 16, no. 3 (1982): 479–504.

Gerrard, Christopher D., *Promoting Third World Agriculture: Lessons of Recent Experience*. Ottawa: North-South Institute, 1983.

Immerwahr, Daniel. "The Politics of Architecture and Urbanism in Postcolonial Lagos, 1960–1986." *Journal of African Cultural Studies* 19, no. 2 (December 2007): 165–186.

John-Alder, Kathleen. "Paradise Reconsidered: The Early Design History of Pradisan Park in Tehran." in Mohammad Gharipour (ed.), *Contemporary Urban Landscapes of the Middle East*, 120–148. London: Routledge, 2016.

Kahama, C.G., William Conklin and James Rossant. *Urban Design for the National Capital Centre*. Dodoma: Capital Development Authority of Tanzania, 1980.

Mitchell, Timothy. "The Object of Development: America's Egypt." in Jonathan Crush (ed.), *Power of Development*, 129–157. London, New York: Routledge, 1995.

The Federal Capital Development Authority. *The Master Plan for Abuja, the New Federal Capital of Nigeria*. Lagos: The Federal Capital Development Authority, 1979.

Scott, James. *Seeing Like a State*. New Haven, CT, London: Yale University Press, 1998.

Shao, John. "The Villagization Program and the Disruption of the Ecological Balance in Tanzania." *Canadian Journal of African Studies* 20, no. 2 (1986): 219–239.

Spirn, Anne Whiston. "Ian McHarg, Landscape Architecture, and Environmentalism: Ideas and Methods in Context." in Michel Conan (ed.), *Environmentalism in Landscape Architecture*, 97–114. Washington, DC: Dumbarton Oaks Research Library and Collection, 2000.

Strother, Zoe. "Architecture Against the State: The Virtues of Impermanence in the Kibulu of Eastern Pende Chiefs in Central Africa." *Journal of the Society of Architectural Historians* 63, no. 3 (September 2004): 272–295.

UN Technical Assistance Administration. *Metropolitan Lagos: Prepared for the Government of Nigeria*. New York: United Nations, Commissioner for Technical Assistance, Dept. of Economic and Social Affairs, 1964.

Vale, Lawrence. *Architecture, Power, and National Identity*. New York, London: Routledge, 2008.

Yoon, Duncan. "The Rationalization of Space and Time: Dodoma and Socialist Modernity." *Ufahamu: A Journal of African Studies* 36, no. 2 (2011): 1–27.

9 Grey, Green, Gold

Uriel Orlow

Uriel Orlow is an artist who lives and works in London and Zurich. His practice is research-based and emphasizes artistic *processes*. His work takes various forms using different mediums, including film, photography, drawing, lecture performances and modular, multi-media installations that focus upon specific locations and micro-histories, bringing different image regimes and narrative modes into correspondence. Orlow is particularly concerned with the spatial manifestations of memory, blind spots of representation and forms of haunting.

The series of works developed under the title *Theatrum Botanicum*, look at the circulation and representation of plants as a revealing platform of socio-political relations. With a specific focus on South Africa and Europe, his enquiry highlights the ways in which the botanical world offers an entry point from which to observe the complex *mise-en-scène* of colonial order. *The Fairest Heritage* for instance (single channel video, 4.18, 19), is based on a series of 1963 films made to celebrate the 50th anniversary of the founding of Kirstenbosch, one of the national botanical gardens of South Africa in Cape Town. The original films, found in the archives of the garden, show a predominantly white society circulating within the perceived neutral world of flowers and plants. Orlow collaborated with the actor Lindiwe Matshikiza to produce double exposed images, in which Matshikiza comes to inhabit the archival film, thereby confronting its naturalization of a historical and politically charged territorial appropriation. In *Geraniums Are Never Red*, Orlow explores the confusion surrounding the naming of geraniums across the world. The plant known today as geranium, was brought back to Europe by the Dutch East India Company from the Cape in the 17th century, and is botanically speaking a Pelargonium. Geraniums correspond to another genus of plant entirely. The current use of the name "geranium" to describe the red plant results from a historical misidentification. The history behind the naming of the geranium thus sets up an enquiry into layered socio-natures and their histories, unveiling the modern myths which continue to inform our present. The following essay by Orlow accompanies one of his works entitled *Grey, Green, Gold*, which explores the interconnected histories of the flower known colloquially since the 1990s as *Mandela's Gold*, or by botanists as yellow Strelitzia Regina, and the history of Mandela's incarceration on Robben Island and the garden he created there with his fellow prisoners. In the essay that follows, Orlow traces the parallels between the *artifices* of colonial plant breeding and colonial management of the political prisoners at Robben Island.[1]

(the editors)

Grey Green Gold

The history of Europeans in South Africa begins with a garden and Vitamin C deficiency. The Cape of Good Hope was midway between Europe and the East Indies where the Dutch East India company was managing a significant colony and regular trade routes.

So it seemed sensible for the Dutch to establish a station on the cape coast to replenish their food supplies—especially water, fresh meat, and fruit and vegetables to counter scurvy, a result of Vitamin C deficiency which left the ships short-handed.

On the evening of April 7, 1652, van Riebeeck a Dutch colonial administrator went ashore at the Cape and two days later he marked out a site for the settlement fort. He lost no time in establishing a garden alongside living quarters and defenses against the local Khoi and San populations.

The garden became known as the Company's garden and still exists today.

—

Strelitzia reginae, so-called "bird of paradise" flowers are usually orange.

They are a stemless, evergreen clump-forming perennial. Greyish green, banana-like leaves grow to a height of about 1.5 m and during winter and spring the large bird-like flowers are held above the foliage on the tips of long, sturdy stalks. The structure and pollination of the flowers is fascinating. The hard, beak-like sheath from which the flower emerges, is called the spathe. This is held at right angles to the stem, and has the appearance of a bird's head. Each spathe contains four to six flowers, and these emerge one at a time from the spathe. Each flower consists of three clear yellow sepals and three deep purple petals. The yellow sepals give the appearance of a crest on the 'bird's' head. Two of the purple petals are joined together around the stamens and the style to form an arrow-like structure. The third purple petal is visible as a small scale.

Nectar is produced by a gland at the base of the flower. In fact, so much nectar is produced that it leaks out and drips down the side of the spathe. Although it is not certain exactly which bird is the pollinator, when it comes to feed on the nectar, it perches on the arrow-like purple petals, which open to expose the anthers and the sticky pollen adheres to the feet of the bird. The bird then transfers this pollen to the stigma of the next flower it visits.

—

In 1961, after ten years of defiance campaigns and several arrests, the Communist Party and the ANC established underground headquarters at a farm in the Johannesburg suburb of Rivonia. Nelson Mandela moved there and evaded security police by masquerading as a gardener and called himself David Motsamyi, which means "the walker."

At the farm the military arm Umkhonto we Sizwe (The Spear of the Nation), was conceived.

On July 11, 1963 leaders of the ANC and Communist Party were arrested at the farm and documents relating to Umkhonto we Siwze discovered.

The ensuing Rivonia Trial took place between 1963 and 1964, and led to the imprisonment of Nelson Mandela and the others among the accused that were convicted on life sentences.

Mandela and his co-accused were transferred to the prison on Robben Island, a small rocky island just off the coast of Cape Town. They remained there for the next 18 years (Figure 9.1).

—

Around the same time that Mandela was sent to Robben Island, yellow-flowering strelitzias cropped up at Kirstenbosch, the National Botanical Garden in Cape Town.

Left to their own devices they will most likely be pollinated by the orange plant and have seeds that breed orange flowers.

To get yellow progeny, two yellow plants must be crossed.

At Kirstenbosch there were seven yellow plants in the nursery.

It took nearly 20 years (more or less the duration of Mandela's stay on Robben Island) of careful selection and hand pollination to build up stock of the yellow-flowering Strelitzia reginae.

Hand pollination is done by mimicking the action of the bird, i.e. using a stick or a finger to depress the purple petals to expose the anthers, scraping some pollen off and transferring it to a receptive (shiny and sticky) stigma on another plant. After fertilization, the fruits develop inside the spathe. As they ripen, the capsules swell and protrude from the spathe, change from green, to brown and woody, and split to release the distinctive shiny black seeds with their tuft of oily orange hairs (the aril) (Figure 9.2).

—

first we had to humanize the prison officers
the warders
they were not accustomed to political prisoners

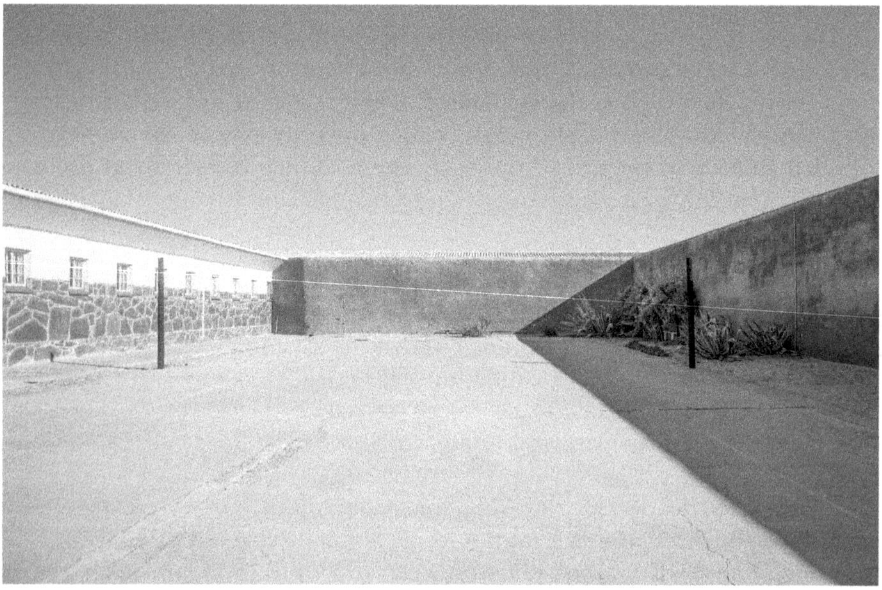

Figure 9.1 Uriel Orlow, *Grey, Green, Gold*, 2015
courtesy of the artist, © 2017, ProLitteris, Zurich

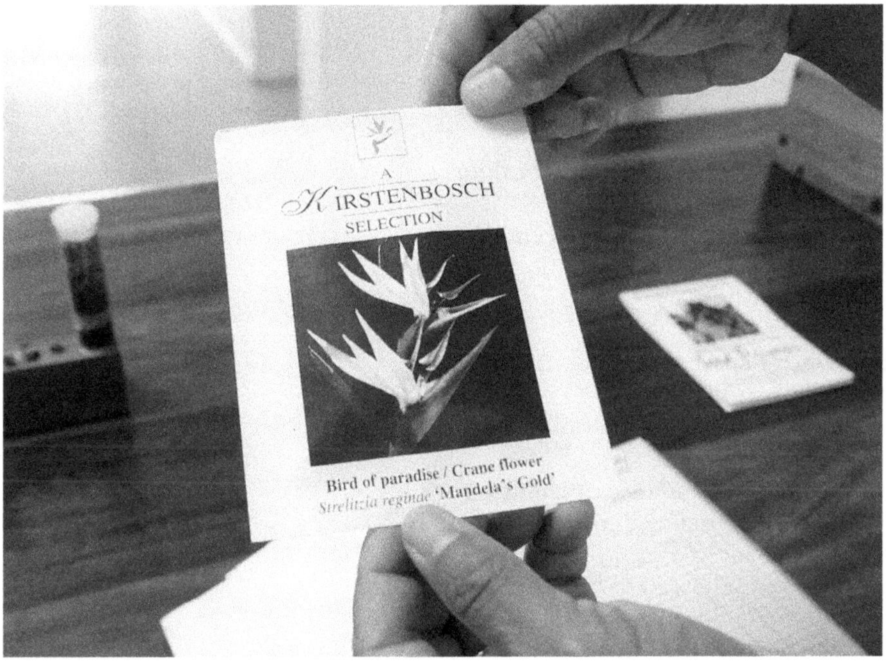

Figure 9.2 Uriel Orlow, *Grey, Green, Gold*, 2015
courtesy of the artist, © 2017, ProLitteris, Zurich

it must have been around six or seven years after we arrived
when they allowed us to have a garden
it started in a very informal manner
some warders were sympathetic to our situation
and one day
one day, one of the warders brought some tomato seeds
we said: fine
let's find a little patch
there was a piece of ground in our courtyard
where we had been knapping stones
so we started using it to grow the tomatoes
and after a little while
somebody brought cucumber seeds
somebody brought melon seeds
somebody meaning the warders
of course plants don't thrive on a barren land
you've got to have fertilizer
there happened to be many ostriches on the island
and their droppings dried in the sun
so when we worked, we picked up the droppings
and the garden flourished

now every Sunday
our religious needs were catered for
and on one occasion
a Hindu priest visited us
after the sermon was finished
he dropped
he dropped a little parcel on the floor
we found some chillies in it
now what do we do with red chilli?
we put them in the sun
surreptitiously
to dry them out
and when the seeds are nice and dry
we will plant them
and the chillies
they grew
there was an abundance of chillies
we asked the prison clinic
to get olive oil
and we were allowed to buy certain groceries
so we bought cocoa tins with an orange lid
and jam jars with a little cap
we harvested the chillies
we filled the jam jars with olive oil
and made pickles
every time we had lousy food in prison
most of the time
we improved the quality of our food

later on
it was coming up to Madiba's sixtieth birthday
we decided
we should make a political statement
it should be some sort of biography
he must start writing
and he did
in the end he wrote six hundred pages
but to smuggle it out
we had to reduce it
reduce it to sixty pages
sixty pages of tiny handwriting
smuggled out of prison
the original manuscript was stored in the cocoa tins
the ones with orange lids
and buried in the garden
the warders eventually found it
and we were punished

the garden
the garden was a strip
between the A and C sections of the prison
from one end of the wall to the other end
about two metres wide
the length of B section
the width of the courtyard

In 1994 the original stock of yellow Strelitzias had been built up enough to be introduced to horticulture. It was registered, released and traded under the name 'Kirstenbosch Gold' until 1996 when the national botanical institute was granted permission to opportunistically re-name it in honor of Nelson Mandela: Mandela's Gold.

There is one problem: at Kirstenbosch, the national botanical garden, where Mandela's Gold is bred, the flower has a predator: The grey squirrel that the colonialist Cecil Rhodes brought with him from England, *Sciurus*, which has naturalized itself there, will consume nearly all the crop if left unchecked. It takes the entire, almost-ripe green capsule, leaving nothing but the vandalized remains of the spathe behind. In order to protect the developing seed each fertilized flower is encaged in fine-mesh chicken wire to keep the European predator, the squirrels out (Figure 9.3).

In 1977 the South African prison authorities allowed a number of journalists to visit the notorious Robben Island. The intention was to persuade the outside world that the conditions there were not as bad as widely believed. On their tour of the island the journalists encountered a tall, thin man dressed neatly in prison clothes and leaning on a spade. The expression on his face was one of intense hostility, and his bearing was

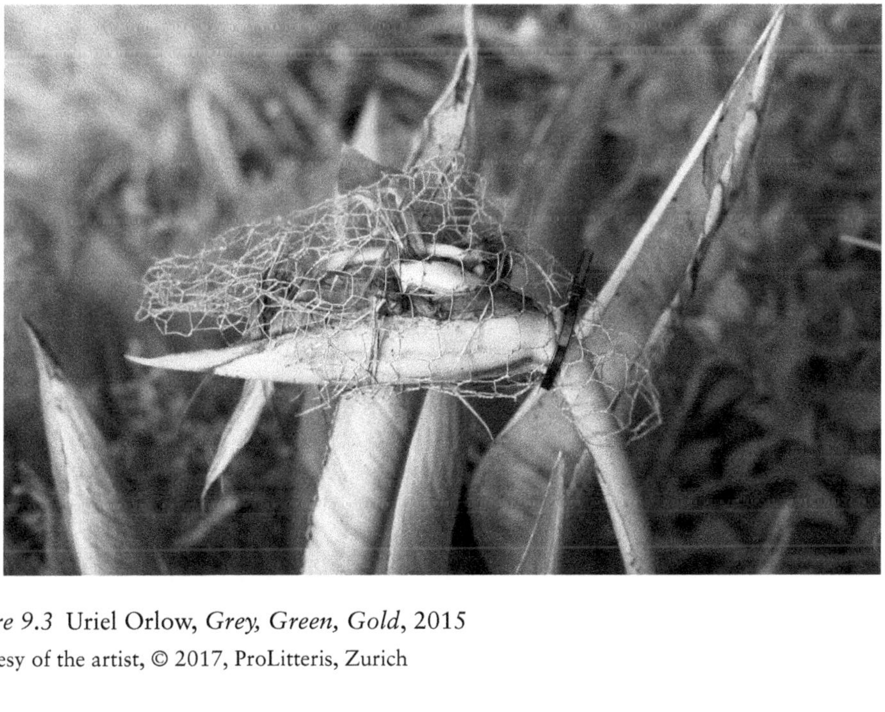

Figure 9.3 Uriel Orlow, *Grey, Green, Gold*, 2015

more that of prince than prisoner. The man was Nelson Mandela, in his 13th year of incarceration on Robben Island.

The photograph was captioned 'A Prisoner Working in the Garden' by the prison authorities. This was far from the truth, as Mandela and his fellow prisoners were forced to do the hard labor of cutting stones on the island.

Mandela immediately set out to draft a letter of protest against the photograph to the prison authorities which was signed by all his fellow prisoners.

It starts like this:

> We strongly protest against the purpose for and manner in which the visit to this prison of the local and overseas press and television men on the 25th April was organised and conducted by the Department of Prisons. We resent the deliberate violation of our right of privacy by taking our photographs without our permission, and regard this as concrete evidence of the contempt with which the Department continues to treat us.

The photograph of Mandela dressed up as a gardener is the only photograph of Mandela from his 18 years in Robben Island prison.

Note

1 The interlocutors for this piece were Laloo Chiba and Ahmed Kathrada, both fellow prisoners of Mandela's at Robben Island. www.urielorlow.net

Part III

Interstitial Spaces

10 Reyner Banham's Desert Landscapes[1]

Eliana Sousa Santos

> "The man who changed the face of America had a gizmo, a gadget, a gimmick—in his hand, in his back pocket, across the saddle, on his hip, in the trailer, round his neck, on his head, deep in the hardened silo."[2]
>
> ". . . the surest way to show you have a mind is to change it"
>
> Reyner Banham[3]

Introduction: From Technology to Aesthetics

Questions of disciplinary focus in architecture are often presented through scale. That is, the expanded field of the discipline is usually defined between the smaller scale of object design and the larger scale of territorial planning, or even the expansion to outer space.

This chapter will present one perspective of the discipline's expansion providing a historiography of architecture from the 1960s to 1980s through an analysis of the work of the English architectural historian and critic Reyner Banham,[4] whose work spans this period. An engineer-turned-historian, in "1960: Stocktaking," a series of articles published in the *Architectural Review*, Banham proposed to blend the arts and the sciences in architecture, although the sciences would feature prominently. Banham's *expanded field*[5] was not so much an expansion of the field of architecture but a hijacking of it by science and technology—for he believed that architecture would have to be invaded by a myriad of technologies in order to maintain its relevance in the future. He redefined architecture as a field that incorporated not only buildings and materials, but also anything that involved creating "fit environments for human activity."[6] In his later work on the American desert, published in the 1980s, Banham merged his previous interests about the design of *fit* and *temperate environments* with aesthetic contemplation.

If in the 1960s his penchant for human ingenuity had led him to study the new and different ways by which humans controlled their surroundings,[7] in the 1980s Banham's travels to the desert, a very adverse environment, showed him the remains of many failed attempts to create *fit environments* in the midst of adverse conditions. The desert seemed to resist, sooner or later, any kind of *technocentric* occupation. The inhabitants of the desert—the 'desert freaks'—fitted their lives to the surrounding environment, and not the other way around. In his text about Solar One, a solar power plant,[8] Banham predicts that soon those attempts to create energy would become obsolete, and rather than discard them, the most sensible thing to do would be to admire them, as the ruins of humankind's good intentions.

This chapter will take the reader on a journey through Banham's writings, from 1960 to 1985, and thus it will look at landscapes and the technological ambitions of the early 1960s, the energy crisis of the 1970s and the shift in policies that occurred in the 1980s.

Stocktaking Architecture: Technocratic Environmentalism

In 1960, Banham edited a series of articles in the *Architectural Review* [AR] called 'Stocktaking Architecture,' which he inaugurated with the essay "Tradition/Technology." This essay is constituted by two parallel texts, one addressing the state of architecture in terms of tradition and the other in terms of technology. This double essay can be seen as an example of his writing model, Banham's texts are compelling because they usually tell a story and present a conflict, sometimes between *two factions*.

In the case of *tradition*, Banham rendered the pursuit of narrowing and focusing the field of architecture. In the case of *technology*, he described the desirable invasion of architecture by professionals from other fields. The double essay proposed the pursuit of the development of *other architecture*,[9] based on the functionalist precept of creating *fit environments for human activities*.[10] This was a deliberately expanded definition of architecture so that architects could operate in other fields and other professionals could invade the architectural field. Banham believed that a main goal of architecture should be to design different environments in the world or, better yet, to design most aspects of human life. He believed, the strict integrity of the architectural profession was undermining the power of the architects in a time when development was deeply connected to innovation.[11] Architecture, for Banham was under a threat of becoming secondary, a background for human activity, while other technologies were becoming essential for human life.

Banham was fascinated with the possibilities of environmental control through the use of devices and *gizmos*.[12] Banham's fascination with the technological element in architecture and the history of its cultural impacts, allowed him to write *The Architecture of the Well-Tempered Environment* (1969). In this book Banham wrote a history of architecture as the production of *fit environments*, as a way of controlling traditionally uncontrollable natural elements. He acknowledged that most of the devices that allow human beings to live in harsh environments are essential to architecture: "the deployment of technical resources and social organizations, in order to control the immediate environment: to produce dryness in rainstorm, heat in winter, chill in summer."[13] Although Banham was critical of the production of bleak functional buildings, his book, at the time, was taken as a threat to the growing endorsement of passive systems.[14] In the second edition of the book, published in 1984, ten years after the oil embargo crisis and the spread of global environmental awareness, Banham wrote to all its detractors, and tried to justify that the book was simply published at the wrong time in history—"the period of growing concern about the apparently irreversible depletion of Earth's energy resources and the pollution of the biosphere."[15]

Between "Stocktaking" (1960) and *The Architecture of the Well-Tempered Environment* (1968) Banham explored the role of architecture as a discipline for controlling the environment, and the work of the architect as an *environmentalist*. In the essay "The Environmentalist"[16] Banham tried to shift the title of *pioneer of the modern movement*[17] to Joseph Paxton. In trying to write a history of modern architecture with Buckminster Fuller as protagonist, instead of Walter Gropius: "in the long haul, the dismissive tag about 'regency greenhouses' does not push Paxton back into the past, but hurls him forward into the Dymaxion world (to coin a phrase) of Buckminster Fuller."[18] This similarity was formal and methodological, Banham wanted to focus upon the architect who would finally practice *other architecture*. To be fully able to design *fit environments* architects would have to be *environmentalists*[19] and Paxton was the model for *The Environmentalist*.

Banham's definition of *environmentalist* included the capacity to change the environment, to transform it so that it becomes *well-tempered* for human inhabitation. However, soon the definition of *environmentalist* was given another meaning. In *Environmentalism* (1970) Timothy O'Riordan identified two approaches regarding the environment: the *ecocentric* and the *technocentric*. The *ecocentric* mode was based on praising "the virtues of reverence, humility, responsibility, and care (. . .) and demands a code of behavior that seeks permanence and stability based upon ecological principles and homeostasis."[20] On the contrary, the *technocentric* mode was based

> in rationality, the *objective* appraisal of means to achieve given goals, by managerial efficiency, the application of organizational and productive techniques that produce the most for the least effort, and by a sense of optimism and faith in the ability of man to control physical, biological and social processes for the benefit of future generation. Progress, efficiency, rationality, and control—these form the ideology of technocentrism that downplays the sense of wonder, reverence and moral obligation that are hallmarks of the ecocentric mode.[21]

Given his interest in the creation of social ecosystems, in 1965 Banham interviewed Roland Wank, the chief architect of the Tennessee Valley Authority [TVA], a large-scale social and environmental project.[22] The TVA was a project that fitted Banham's expanded definition of architecture as the creation of *fit environments* for human activities. The TVA was to serve as a catalyst for social and economic development in a region that was deeply deprived. This was an area prone to natural disasters and deep poverty; piecemeal improvements had been made before, but they were not advantageous to the region. The TVA project developed the region considerably, mostly through the construction of hydroelectric dams. In the 1960s, when Banham interviewed Wank, the TVA was close to the peak of its development. Banham's notes about the interview stress the utopian principles of the TVA project. The design of the dams would reflect their incitement to prosperity.[23] However, the affluence of the region did not grow exponentially, and when Banham published his article about the TVA in 1977,[24] the validity of the project was being questioned. Moreover, the scale and ambition of the project was being taken as a symbol of *technocentric* development, and used as a derogatory term by environmentalist sympathizers.

Banham's *environmentalism* fit O'Riordan's definition of *technocentrism*. In the archive about his travels on the American Desert Banham refers to O'Riordan's book along with Daniel Defoe's Robinson Crusoe.[25] Given this reading material, I dare say that while cycling in the American desert he thought about the balance between *ecocentric* and *technocentric* environmentalism. The separation between these two categories is only a matter of point of view, since they ultimately share common goals.

Banham's Anti-Picturesque

In the essay "The revenge of the picturesque"[26] Banham acknowledged that the *picturesque* campaign—started by the *Architectural Review* and mainly headed by Nikolaus

Pevsner[27]—was absorbed by architects in the late 1950s and 1960s. Banham pointed towards historicism, and found the abuse of naïve references, or easy analogies to complex systems to be one of the consequences of the emergence of the *picturesque*. In this account of the evil consequences of the *picturesque* comeback, Banham characterized Pevsner's work as a sign of a universal regressive change in the history of architecture, which presented the modern movement as the heritage of earlier movements.

Banham's text shows a detailed account of this battle between those who defended the *picturesque* ideals and those against them. But if his story starts with an organized description of two factions where everybody has their well-defined place in either one of the two—leading the reader to think that there would really be a battle between the *good rebels* and the *bad traditionalists*—by the end, these boundaries based on principles are not so clear. For Banham, the boundaries got mixed when *relativity* came to be used as analogy in the general approach to art and design. The historical association that connected German expressionism, *art nouveau*, *baroque*, the *landscape garden* and the *picturesque* made sense for completely novel reasons:

> But much had happened to destroy the congruities of geometrical beauty [. . .] and members of the anti-Picturesque connection who were interested in such topics were already making free with concepts such as Heisenberg's Uncertainty Principle, were growing suspicious of *one-to-one* relationships and the concept of *uniqueness*, were beginning to talk of topology rather than geometry, and if they did not yet dispose of the concept *open-ended*, were certainly reading Professor Karl Popper's The Open Society as implying the downfall of all closed and determinate systems such as Plato's politics—or classical architecture based on elementary geometry.[28]

The resistance towards the gigantic scale of modernist architecture was being felt by the multiplication of complaints about its generic character. References to gradual growth, and concepts such as Popper's *piecemeal social engineering* were part of this tendency. With this, some of the notions that emerged with the revival of the *picturesque* were appropriated, like cherishing the palimpsest quality of cities. There was a call for a different scale of action, and the *picturesque* benefited with this moment.

Banham made the case for the *do-it-yourself* culture that thrived in America. The *pioneers of modern design* in America were not isolated individuals, but those who invested in the collectivization of knowledge. This fascination with the functional and ingenuity was, for Banham, the mark of *other architecture*. Another side of the paradigm of the changing environment was the use of objects and design as catalysts of change. The idea that establishes design as problem solving was "invented in America,"[29] because there was an absolute necessity to occupy a territory as efficiently as possible.

Banham's position, defending the culture of fast colonization that built America in a relatively short time, became problematic in the 1970s. What in the 1950s might have seemed a symbol of Progress (e.g. the massification of culture and design objects that allowed the final separation of man from nature) became nostalgic symbols of an imagined past, where the atrocities of colonization were forgotten. The objects that allowed the 'American pioneer' to conquer the land and to exploit the continent from east to west were now thought as objects that allowed exploitative pursuits. These objects were not *good* objects.[30]

In the 1970s, the global gradual dissatisfaction with the strictness of over-controlled environments along with the connection of streamlined design to corporate and capitalist structures, gave way to desires to explore its opposite. If the streamlined became conventional, the radical would be in the work of the (DIY) *bricoleur*.[31] Banham, would have praised the ingenuity of *do-it-yourself* culture. For Banham, American design was characterized by the use of the object—the *gizmo*—and this use was usually connected to the necessity of changing the environment. The allure of American design resided in the fact that those were tools that could be used by unskilled labor. They were objects that allowed their users to conquer space and establish their lifestyle in harsh lands. The *pioneers* of American design that Banham praised, who he emulated by wearing a Stetson hat, used their *gizmos* and tools to change the environment.

Aesthetics Are the Ethics of the Future

Banham's positioning as part of the anti-Picturesque faction would be contradicted by his later works *Los Angeles: The Architecture of Four Ecologies* (1972) and *Scenes in America Deserta* (1982). His initial fascination with technology would be replaced with a predilection for irregularity, a distaste for the formulaic and fondness for the re-emergence of past memories connected to life or the appreciation of artworks. Banham also developed documentaries about the same subjects that were shown on BBC *Reyner Banham Loves Los Angeles* (1972) and *Roads to Eldorado* (1979).

When Banham wrote his book about Los Angeles, the city was commonly seen as the most prominent example of bad urban planning, an ugly, unfriendly city consumed with smog. Praise for Los Angeles was at that time considered outrageous.[32] Besides the book he also did a film documentary. In both media forms, Banham showed the viewer the city's beauty. The noted qualities of Los Angeles were similar to the qualities that Pevsner identified in the *picturesque*, since the city is described as surprising, irregular and at the same time well known through the myriad of representations in which it is depicted—painting, literature, songs, and obviously, films.

To illustrate the *picturesqueness* of Banham's account of Los Angeles let us take as an example the documentary 'Reyner Banham Loves Los Angeles.'[33] In this film one can see how the influences of different media and Banham's own association of ideas characterized his view of the city. The documentary was set up as a travelog to train the eyes of those who did not yet know how great Los Angeles was. Similarly, Banham's proclaimed love for this city comes from his previous knowledge of the city's representations; he already knew the city long before he went there.

Referencing the well-known tourist guides—the Baedeker—Banham drives around in a car with a device called Baede-Kar, a *gizmo* that helps him explore and explain the city. 'Los Angeles needs some explaining' Why? Because it was considered a sprawling mess—a 1970s version of the 1870s 'let it grow freely' *picturesque* garden design applied to a city. Not only gardens grow and change by themselves, people's lives and choices can transform a city, and Los Angeles was a city where Banham could forget about history and concentrate on the actions of the people he identified as catalysts for the city's change. The city, apparently, without planning, and growing naturally was, for Banham, a present vision of the future.[34]

Why was this a city of the future? Banham tells us that Los Angeles is a city that makes "no sense of history and breaks all the rules." As a historian, Banham operated in a similar way as Charles and Ray Eames, designers who practiced a design of miscellany and inclusion of references. Banham found the Eames's house in the Pacific Palisades to be "the house that taught the world to come to Los Angeles" and that showed that "machine age materials could be beautiful, even pretty and make a proper setting for beautiful objects."[35] The references to multiple cultures embraced by these architects showed their variegated preferences: American popular culture, Indian folk art, Japanese architecture, and futuristic ironies, all of them created a patchwork of interests that appealed to Banham. For him the city of Los Angeles was equally a patchwork of manifold things.

To show the differences between Los Angeles and other cities, Banham returned to England. He goes to London and to Norwich, "the fine city with a cathedral" where Banham was born, and mentions that it was there that he had his first contact with Los Angeles, in the *penny pictures*, where he and his colleagues could be *transported to Los Angeles*. Their image of the city was educated by Buster Keaton's movies. So that with the help of film and television they were "all familiar with the place already."[36] Most of the city's sites shown by Banham are material evidence of ideas, they are sites connected with the mythical, the artistic and the philosophical, and at the same time they feature in popular culture. Along with Hollywood's fabricated memories, Banham saw Los Angeles as a city where different lifestyles and trends converged. He interviews Ed Ruscha and Mike Salisbury as part of the Los Angeles art community, and as contributors to the building of the city's image, especially Ruscha who documented obsessively the city's streets, its gas stations and parking lots. In the film Banham is clearly driven to look at the city through elements extraneous to it, elements that mediate his view—artworks, films, songs and lifestyles.

Similarly, Banham's account of his travels in the American desert can also be considered mediated. Many times when facing objects or views in the American desert Banham immediately associated them with artworks. The book—*Scenes in America Deserta*—is an organized account of a *picturesque* experience. When Banham arrived at the Red Desert he described his emotions as if he were playing a part in a Western film: "The view was an exaltation. I leaned breathless in the wind, grasping my Stetson to my head and, through stinging watery eyes, surveyed the gigantic prospect before me."[37]

Reyner Banham's *Scenes in America Deserta* (1982)

Initially Banham's intention was to travel in the desert so as to find something truly original, nonetheless Banham could not help but admit the myriad of references that informed his views on that place: "This time I had to figure out for myself why, and how, these scenes appeared beautiful to me, since no eye of authority had laid the matter out before me. I had seen no work of art that in any way resembled the great Soda Lake . . . or rather, I had seen such a work of art and wish I hadn't." He had already seen works of art and read accounts about the desert presenting the sites he was now seeing, or trying to see. Banham was even favorable towards the *picturesqueness* of such well-known sites as Route 66, where site specificity was not at all important, just the mythical memories that the place held: "It is still exciting to be on the 'historic Route 66' where it physically survives, to pass through roadside settlements farmed in story and song."[38]

The desert was supposed to be void, it was supposed to be silent, but it was full of *picturesque* or mediated references. In the notes about his travels he notes that he is "In Judd's grounds, having a *thing*"[39] just like *the philosophers of the picturesque* in the 18th century might have had *things* when traveling in a landscape recognizable as a celebrated painting.

Banham wanted to go to a place devoid of all references, but the desert was not such a place. The desert was not a desert in the literal sense. Banham had a reference that depicted the American desert as an abstract entity; the ultimate example of characterization of the American desert was *The Desert* (1901) by the art historian John Van Dyke. For Banham, Van Dyke's book had the particular quality of showing the desert as a place full of abstract visual qualities. Banham wrote a review about a reprint edition of the book where he remarked that the renewed interest in the book might have had the wrong reasons. The reprinting of the *Desert* was made "on currently conventional eco-conservationist grounds, which is a pity because Van Dyke's views on these subjects go little beyond the received wisdom of those Teddy Roosevelt years, while his observation of the desert life must now seem rather less than remarkable." Van Dyke's book was, on the contrary, remarkable by its aesthetic descriptions: "show him a crag at sunset, the haze in a valley or the fall of light on salt or sand and he is incomparable."[40]

Banham was rather impressed with the precision of Van Dyke's descriptions: "What makes the result so compelling is its visual accuracy, and command of a language for describing colour and light. What held one transfixed among the linens was a classic shock of recognition; here, for the first time in literature, was the full range of desert scenery as only one who has been there can know it, rendered verbally with a truthfulness and power to stimulate the eye of imagination which has not been rivaled since."[41]

The precision with which Van Dyke described the colors and textures, of both air and ground in the desert, made a profound mark on Banham as one can note in several descriptions that can be found in his travel notes where the arduous search for the right term is evident and the descriptions of color gradients show a considerable effort: "Sun at left (2.30–3.30) very deep blue/purple shadows at either side of mountains to left. Snow-covered peaks above dark brown mesas to right."[42]

Finally, it was Van Dyke—rather than Hume or Payne Knight—who taught Banham that: "The rewards of the desert are in the eye of the beholder."[43] Richard Payne Knight had used Hume's notion that beauty was to be found in the 'mind of the beholder' as a leitmotiv in his *An Analytical Enquiry to the Principles of Taste* (1808). Banham knew that what he noticed in his trip to the desert were things that, for some reason, he already knew and cherished before. Banham allowed himself to philosophize about the 18th century's aesthetic trinity, he asked himself what the desert might be: *beautiful*, *sublime* or *picturesque*? He concluded that the desert cannot be beautiful because it is too arid; and surely not picturesque, since there were not enough artistic references and only partly sublime since he could not find the sense of threat.

Banham's question is thus very similar to the one Uvedale Price posed two hundred years before. Uvedale Price wrote his *Essay on the Picturesque* in 1794 as an amendment to the famous Edmund Burke's *Inquiry into the Origin of our ideas of the Sublime and the Beautiful* (1756). Price thought that another category of objects should have been added to Burke's classification, those of objects that could not be considered neither *beautiful* nor *sublime*. How can we characterize something that is neither *beautiful* nor *sublime*? Price answered: *picturesque*. Banham, who considered that this word had suffered particular conceptual erosion, tried to see the desert as something other between the concepts of beautiful and sublime:

> Here, as a historian, I must distinguish between 'the beautiful' in its current sloppy usage (which I myself have used above) and in its more precise C18 sense as the

> middle term between 'the Sublime' and 'the Picturesque'. A 'picturesque' view of the deserts I have obviously discarded, since there are not enough authoritative pictures for them to resemble, and 'beautiful' in the Uvedale Price sense of 'soft, rounded feminine' they equally obviously are not. Are they, then, 'sublime' like the Alps, or like the rock of El Capitan in Yosemite? The answer could be a cautious 'yes', though little of the true desert has the majesty, or the sense of threat, that originally informed the sense of the Sublime.[44]

Banham deliberately avoids the word *picturesque* however because of the associations it contained, memories of built *picturesque* revivalism that he mentioned in the essay "The Revenge of the Picturesque" (1968). The desert had to seem devoid of human references. So the desert could not be *picturesque*, because this word carried a sense that was still embedded in Banham's mind.

Robert Smithson, a vocal partisan of the *picturesque* principles,[45] also taught Banham how to look at the desert, which although mythically suggesting an unspoiled nature, was inhabited and transformed, it was *man-mauled*. In a text about the photographs of Richard Misrach, Banham recognizes that the desert suggests many images. The desert is a place that is *picturesque* in that sense. The desert devoid of all references was also one of these images, but that was an image easy to dispute. The premise of *Scenes in America Deserta* (1982) was that the desert is not deserted. For Banham, Robert Smithson's work was an attempt to look at the desert without nostalgia for a *deserted desert* or resentment towards its human intervention. Robert Smithson taught Banham that there is beauty in that intervention:

> In guilt or despair we try to block out these manifestations of our presence and lift up our eyes unto the hills beyond—hoping that they, at least, will still look something like the work of Ansel Adams. They don't of course; the power lines loop across their flanks, the television repeater towers sit in the high passes between them, and the navigation beacons flash from their summits. It is a scene that many of us would rather not see, because we have no proper preparation for seeing it, in spite of the efforts of earthwork artists like Robert Smithson to make us see that some humans, at least, can make the desert better by their interventions.[46]

It is one of these human interventions that is described in "The Most Beautiful Power Plant in the World,"[47] Solar One built in the Mojave Desert in California. This was the first project of a solar power plant to use 1,800 mirrors. In 1960 Banham might have looked at this project as the perfect union between art and engineering, as something that would create a *fit environment*, but in 1985 he was more interested in its aesthetic qualities.

The title is connected to the aesthetic lure of the object, but Banham called it beautiful in three instances: aesthetics, ethics and ingeniousness. All these qualities merged in the "largest and most productive solar energy installation in the world" but could be also a "pretext for making what may turn out to be the biggest desert artwork ever."[48] The visual power of the plant was as powerful as its energy production ability.

Banham was fascinated with this project because it showed the role of ingeniousness in the improvement of the environment: "everybody knows that the Golden Gate bridge is beautiful, but what about Stanford linear accelerator or the Moffett field dirigible hangar?"[49] Aesthetics, ingeniousness and ethics together would set a complete new standard for design.

Solar One was designed and built based on hope when the run for solar energy began in 1974, with the Solar Energy Research Development and Demonstration Act as a direct consequence of the 1973 oil crisis. He describes the site as a "spectacular array of 1818 heliostat mirrors, each 23 feet square, all turning and tilting at about the pace of a clock's hour hand under meticulous computer control that keeps them relentlessly reflecting the sun's light at a boiler that sits on top of a 300 foot tower and flashes super-pure water into steam at nearly 1000 degrees Fahrenheit. The concentration of reflected light on the boiler is such that it glows like a miniature sun itself."[50]

This must have been such a spectacular site that Banham asked himself: "It's terrific, a unique spectacle, but where are the crowds?"[51] He manifested his disappointment with the generic lack of interest about this structure. Part of the problem was that the structure didn't fit the stereotype of passive energy systems; it was a solar power plant more suited to the image of the *space race* of the 1950s than to the return to the simple life of the 1970s: "Passive solar is seen as something done at a modest and domestic level by earnest, if slightly cranky bearded folks in funny-shaped houses. It is popularly expected to be user-friendly, not to say user-cuddly, and that is something that an installation as vast and aloof as Solar One will never be."[52] Banham used a similar argument to defend the TVA project. Both structures, the TVA and Solar One, were ethically inspired projects, however their images were not correlated to the ethical image of passive systems. In this case there was a contradiction between the image and the use of the actual power plant. For Banham some designs might suggest an *ethical* image thorough aesthetics. Whereas Solar One, while being an ethically sound project conveyed an image of *high-tech* aesthetics: "What we are looking at was perhaps the largest accidental work of art ever to stand on the earth's crust."[53]

Banham made his case to change the category of Solar One from a power plant to a work of art, because he was aware of its short life span. This was one of the first attempts to build a solar power plant and had an obvious drawback, because the desert's dusty nature was an obstacle to the power plant's optimal function: dust particles would eventually build deposits on the mirrors and obstruct the reflection of light, so that every mirror had to be hand washed every once in a while. This meant that the useful life of this beautiful object was limited, therefore Banham decided to classify it as a work of art so that its conservation could be made possible: "it becomes a work of art only when it has stopped performing and gone into its resting configuration."[54] Ironically, the historian of the immediate future concluded his praise for this object with a call for its conservation: the usefulness of technology had been outlived but the beauty of form would remain.

Notes

1 This essay is co-financed by the POPH program of the European Social Fund and by National Funds by FCT Foundation for Science and Technology, with a postdoctoral grant with the reference SFRH/BPD/68962/2010.

2 Reyner Banham, "The Great Gizmo," in *A Critic Writes: Essays by Reyner Banham, 109–18*. (Berkeley: University of California Press, 1999), 109.

3 Reyner Banham, *Guide to Modern Architecture*. (London: The Architectural Press, 1962), 11.

4 Peter Reyner Banham (Norwich 1922–London 1988) first trained as an engineer became an architectural historian and critic, after having studied under Nikolaus Pevsner (Leipzig 1902–London 1983). Banham was a prolific writer and broadcaster, his doctoral thesis *Theory and Design in the First Machine Age* (1960) became a reference to the history of architecture in the 20th century, following his supervisor's *Pioneers of the Modern Movement: From William Morris to Walter Gropius* (1936). For a detailed account on Banham's

biography see Nigel Whiteley, *Reyner Banham: Historian of the Immediate Future.* (Cambridge, MA: MIT Press, 2002).

5 Nigel Whiteley borrowed Rosalind Krauss's expression to characterize Banham's search of *other architecture* and *fit environments for human activities* in Whiteley, *Reyner Banham: Historian of the Immediate Future*, 187–243.

6 Banham, Reyner. "1960: Stocktaking," *Architectural Review*, 127, no. 756 (1960), 93.

7 Reyner Banham, *The Architecture of the Well-Tempered Environment* (Chicago: University of Chicago Press, 1984).

8 Reyner Banham. "The Most Beautiful Powerplant in the World." *California* (November 1985), 98–103.

9 The term 'une architecture autre' or *other architecture* is central in Banham's work, that first appeared in connection to his definition of Brutalist architecture as related to Art Brut's search for 'un Art Autre'. See Whiteley, *Reyner Banham: Historian of the Immediate Future*, 130–131.

10 Banham, "Stocktaking," 39.

11 Ibid., 94.

12 Banham wrote the essay "The Great Gizmo" in 1965 where he makes the argument that gadgets changed the landscape and the territory of America, see Banham, "The Great Gizmo."

13 Banham, *The Architecture of Well-Tempered Environment*, 18.

14 In a review about Banham's book, other books are considered as giving a better view about the relation between architecture and environmental control, such as Bernard Rudofsky's *Architecture without Architects* (1964) and Victor Olgyay's *Design with Climate* (1963) in James Marston Fitch. "Review," *Journal of the Society of Architectural Historians* 29, no. 3 (1970), 282–284.

15 Banham, *The Architecture of Well-Tempered Environment*, 13.

16 Reyner Banham, "The Environmentalist," *Program: Columbia University* (Spring 1962), 58–64.

17 In his seminal work *The Pioneers of Modern Movement: From William Morris to Walter Gropius* Nikolaus Pevsner describes a lineage started by William Morris and the *Arts and Crafts* movement.

18 Ibid.

19 Later Banham used the definition of *environmentalist* to characterize Frank Lloyd Wright in Reyner Banham, "Frank Lloyd Wright as Environmentalist," *Arts & Architecture*, no. 83 (1962), 26–30, and mentions Joseph Paxton as an exemplary designer of *machines à habiter* in *The Architecture of Well Temperate Environment*, 144–145.

20 Timothy, O 'riordan. *Environmentalism* (London: Pion Limited, 1981), 1.

21 Ibid., 11.

22 This research resulted in two essays, that were published later (Reyner Banham, "Valley of the Dams." In *A Critic Writes: Essays by Reyner Banham* (Berkeley: University of California Press, 19990. and Reyner Banham. "TVA: The Engineering of Utopia," *Casabella*, no. 542–543 (1988): 74–81), in Banham's archive there are notes of the interview he conducted in Banham, Reyner. "Interview with Roland Wank, TVA's First Architect," 1965. Reyner Banham Papers Box 12. Getty Research Institute, folder 10.

23 Banham, "Interview with Roland Wank, folder 10.

24 Banham, "Valley of the Dams."

25 Reyner Banham, "Scenes in America Deserta Manuscripts." Los Angeles, nd. Reyner Banham Papers Box 1:2. Getty Research Institute, folder 2.

26 Reyner Banham, "The Revenge of the Picturesque," in *Concerning Architecture: Essays on Architectural Writers and Writing Presented to Nikolaus Pevsner* ed. John Summerson (London: Penguin Press, 1968), 265–273.

27 By the end of the second World War the editorial team of the AR, under the direction of Hubert de Cronin-Hastings, launched a rescue of the *picturesque*, a word that was already part of a vocabulary of abuse—partially inspired by Christopher Hussey's *The Picturesque: Studies in a Point of View* (1927)—to justify Gordon Cullen's theory of visual planning, *townscape*. Each issue of the AR would include essays about the historical context of the *picturesque*—mostly written by Nikolaus Pevsner, Cronin-Hastings and their pseudonyms—to be read in

parallel with Cullen's visual essays. See Vidler, Anthony. *Histories of the Immediate Present: Inventing Architectural Modernism.* (Cambridge, Ma: MIT Press, 2008).

28 Ibid., 265.

29 Reyner Banham. "Europe and American Design," in *Lessons From America*, ed. Richard Rose (London: Macmillan, 1974), 67–91, 80.

30 This connotation still lingers in the present in arguments about American political instances. The *gun*, the *truck*, the *cowboy hat* and so forth, objects that are used to resist the adversity of the environment became the symbols of conservative America.

31 This refers to Claude Lévi-Strauss's reference to the ahistorical savage mind—the mind that was more prone to the concrete, the mythical, *bricolage*, intuition and atemporality, than to the abstract, the scientific, *engineering*, abstraction and history—became a model and an analogy to the critique of generic history. See Claude Lévi-Strauss, *The Savage Mind* (London: Weidenfeld & Nicolson, 1972).

32 This is exposed in Anthony Vidler. "Los Angeles: City of the Immediate Future," in *Los Angeles: The Architecture of Four Ecologies* (Berkeley: University of California Press, 2001), xvii–xxxiii.

33 Julian Cooper, "Reyner Banham Loves Los Angeles," *One Pair of Eyes*, 1972. http://video.google.com/videoplay?docid=1524953392810656786#.

34 Ibid.

35 Ibid.

36 Ibid.

37 Reyner Banham, *Scenes in America Deserta* (London: Thames & Hudson, 1982), 50.

38 Reyner Banham, "Scenes in America Deserta Manuscripts." nd. Folder 2.

39 Ibid.

40 Ibid.

41 Ibid.

42 Ibid.

43 Ibid.

44 Ibid.

45 In 1973, Robert Smithson published an essay that allowed people to look at his own work as *picturesque*, "Frederick Law Olmsted and the Picturesque Landscape." Inspired by his visit to the exhibition about Frederick Law Olmsted's work in New York's Central Park, Robert Smithson referred Olmsted's interest in the writings of Uvedale Price and William Gilpin: "The origins of Olmsted's view of landscape are to be found in 18th-century England, particularly in the theories of Uvedale Price and William Gilpin. Price extended Edmund Burke's Inquiry into the ideas of the Sublime and the Beautiful to a point that tried to free landscape from the *picture* gardens of Italy into a more physical sense of the temporal landscape. A tree for example, stuck by a lightning was something other than merely beautiful or sublime—it was *picturesque*." Robert Smithson, "Frederick Law Olmsted and the Dialectical Landscape," *Artforum* 11, no. 6 (1973), 62–68.

46 Banham, *Scenes in America Deserta*, 1.

47 Banham,"The Most Beautiful Powerplant in the World."

48 Ibid., 98.

49 Ibid., 100.

50 Ibid., 102.

51 Ibid.

52 Ibid.

53 Ibid., 103.

54 Ibid.

References

Banham, Reyner. "Scenes in America Deserta Manuscripts." Los Angeles, n.d. Reyner Banham Papers Box 1:2. Getty Research Institute, folder 2.

Banham, Reyner. "1960: Stocktaking." *Architectural Review* 127, no. 756 (1960): 93.

Banham, Reyner. "Frank Lloyd Wright as Environmentalist." *Arts & Architecture* no. 83 (1962): 26–30.
Banham, Reyner. *Guide to Modern Architecture*. London: The Architectural Press, 1962.
Banham, Reyner. "The Environmentalist." *Program: Columbia University* (Spring 1962): 58–64.
Banham, Reyner. "Interview With Roland Wank, TVA's First Architect," 1965. Reyner Banham Papers Box 12. Getty Research Institute, folder 10.
Banham, Reyner. "The Revenge of the Picturesque." in John Summerson (ed.), *Concerning Architecture: Essays on Architectural Writers and Writing Presented to Nikolaus Pevsner*, 265–273. London: Penguin Press, 1968.
Banham, Reyner. "Europe and American Design." in Richard Rose (ed.), *Lessons From America*, 67–91, 80. London: Macmillan, 1974.
Banham, Reyner. *Scenes in America Deserta*. London: Thames & Hudson, 1982.
Banham, Reyner. *The Architecture of the Well-Tempered Environment*. Chicago: University of Chicago Press, 1984.
Banham, Reyner. "The Most Beautiful Powerplant in the World." *California* (November 1985): 98–103.
Banham, Reyner. "TVA: The Engineering of Utopia." *Casabella* no. 542–543 (1988): 74–81.
Banham, Reyner. "Valley of the Dams." in *A Critic Writes: Essays by Reyner Banham*. Berkeley: University of California Press, 1999.
Banham, Reyner. "The Great Gizmo." in *A Critic Writes: Essays by Reyner Banham*, 109–118. Berkeley: University of California Press, 1999.
Cooper, Julian. "Reyner Banham Loves Los Angeles." *One Pair of Eyes*, 1972. Online video. retrieved on January 2010 from http://video.google.com/videoplay?docid=1524953392810656786#.
Fitch, James Marston. "Review." *Journal of the Society of Architectural Historians* 29, no. 3 (1970): 282–284.
Lévi-Strauss, Claude. *The Savage Mind*. London: Weidenfeld & Nicolson, 1972.
O 'riordan, Timothy. *Environmentalism*. London: Pion Limited, 1981.
Smithson, Robert. "Frederick Law Olmsted and the Dialectical Landscape." *Artforum* 11, no. 6 (1973): 62–68.
Vidler, Anthony. "Los Angeles: City of the Immediate Future." in *Los Angeles: The Architecture of Four Ecologies*, xvii–xxxiii. Berkeley: University of California Press, 2001.
Vidler, Anthony. Histories of the Immediate Present: Inventing Architectural Modernism. Cambridge, MA: MIT Press, 2008.
Whiteley, Nigel. *Reyner Banham: Historian of the Immediate Future*. Cambridge, MA: MIT Press, 2002.

11 'Nature' in Interstice

Reflections on China's "Obscure Poetry" in the 1970s

Jundan Zhang

> We need and are perhaps beginning to find different ideas, different feelings, if we are to know nature as varied and variable nature, as the changing conditions of a human world.
>
> —Raymond Williams, 1980, Ideas of nature, p. 85

The 1970s in China was an obscure period. While Chinese society underwent dramatic social, political and economic transitions during the 1960s and 1980s, the 1970s appeared only to be a *necessary* temporal connection. As Li Tuo, a Chinese writer, describes: "These ten years seemed to be hasty and swift, like an interlude between two monumental performances, or an interstice in the historical development."[1] However, it is exactly this interstitial moment that offers a unique perspective into many crucial issues. In this chapter, I explore the problem of the representations of nature in China within the temporal interstice of the 1970s. China has been in the international spotlight due to worrisome environmental damage and its environmental strategies/ policies since the 1978 economic reform. However, as Tong points out, the picture of China's environmental issues is both a fact and a myth.[2] Although environmental damage is real, few people have a proper understanding of the historical background behind it. Therefore, just as Li Tuo questioned the "human" in his seminal anthology "The Seventies," this essay questions "nature" in that same period, exploring how literature of that period might help us understand the shifting representations of nature in China; and exploring how such shifts influenced discourses on 'nature' in contemporary China. Borrowing Li Tuo's word "interstice," I ask what did 'nature' look like in the 'interstice' of the 1970s? Within a reflection on the representations of nature, the "interstice" not only stands as a physical gap and a period between times and spaces, but also as a filter through which we are engaged with different representations. "Interstice" thus captures the sense of historical temporality and intellectual ambiguity in a moment of meaning-shifting and representation-transiting.

Modern Chinese literature provides a unique resource to explore this interstice. There has been a growing "green dimension" in modern Chinese literature. Beginning in the 1980s, but becoming more visible in the last decade, a burgeoning ecological awareness in Chinese literature has shifted the focus to "the shaded side of contemporary China."[3] While many scholars within and outside of China have examined these texts (produced since the 1980s) from an ecological and ecocritical perspective, there has been little attention to the literature of the 1970s. This may be due to the fact that the literature scene in the 1970s was so closely intertwined with issues of political instability and social reform that scholars have tended to focus on those aspects, and neglect ecological issues/themes. It is also true that the various literary forms of the 1970s were still struggling to survive, and at the same time evolve in the aftermath of the Cultural Revolution. However, one unofficial poetry form known as "Obscure Poetry" can offer a starting point to investigate the neglected decade of 1970s China.

Obscure Poetry in Chinese Modern Literature

On December 23, 1978, the first non-official literary journal in China since 1949, *Today*, was launched. It is through this journal that a group of poets' works were published and later labeled as "Obscure/Misty Poetry." In order to understand the position of "Obscure Poetry" in history, we need first to situate *Today* and its 'non-official' modes of disseminating literature in the context of Chinese modern literature of the 1970s. It will be shown that Obscure Poetry was incubated and developed through a series of interstices in a highly repressive society and thus challenging the mainstream/official literature forms.

Chinese society in the 1960s was highly and homogeneously politicized, and literature was almost exclusively used as political eulogy. Even so, there were small groups of intellectuals searching for interstices and hidden spaces to cultivate and express their own voices.[4] These intellectuals (both men and women) were often in their twenties with political and military elite family backgrounds. Their poetry is known as "pre-Obscure Poetry" and is seen as an early trace of using poetry as a means of expressing political protest and individual identities.[5] During the period of the Cultural Revolution, a cultural vacuum in literature and the arts was created after the removal of the old cultural leadership in the late 1960s.[6] A large number of educated youth were deported to the countryside; there they also enjoyed a certain liberty in the countless 'interstices' created by the remoteness from the political center. In the third and last phase of the Cultural Revolution, from 1971 to 1976, with several major events (the deaths of several party leaders including Mao and the end of the Cultural Revolution) affecting the political climate, a more vibrant and alternative thinking blossomed in these interstices and several tendencies of poetry emerged which quickly filled in the cultural vacuum that had prevailed.[7]

Some of these tendencies were recognized as being more (officially) in line with the dominant political ideologies and thus became supported. Provincial and city literary journals resumed publication in 1972. In January 1976, the historic cultural magazines *Poetry* and *People's Literature* revived. However, literary works in these official literature magazines were rather uncritical.[8] Meanwhile, the young intellectuals who used poetry as a means of political and personal expression refused to become the so-called 'official writers' and started to form their own 'non-official' platform.[9] *Today* was built as an alternative platform for unorthodox thoughts and to promote the work of young intellectuals. A group of young poets gathered in *Today* and published poems that would later be dismissively labeled "Obscure Poetry" by Zhang Ming an 'official' writer from the *Poetry* magazine. Besides Mang Ke and Beidao, the two major editors of *Today*, the other important poets who contributed regularly to the journal included Gu Cheng, Yang Liang, Shu Ting and Jiang He.

An Ecocritical Rereading of "Obscure Poetry"

Today lasted only two years and was banned in September 1980. However, the influence of "Obscure Poetry" remained and it is believed that "all poetry of significance written in Mainland China over the past fifteen years or so has derived from or developed in conscious relation to Misty (Obscure) Poetry."[10] Beidao, one of the editors of *Today*, also stated that "The 'misty' or 'obscure' poetry . . . was able to challenge the dominance of the official social discourse, by opening a new space and possibilities for the modern Chinese language."[11] Meanwhile, "Obscure Poetry" reached out to an international audience in an unprecedented manner. "Obscure Poetry" also drew academic attention, but critical analysis has mostly focused on political content, while significant themes such as love, friendship, nature and history have been much less discussed.[12]

An ecocritical approach offers an entry point to understand the influence of "Obscure Poetry" within the wider environmental consciousness and discourse in contemporary China. Ecocriticism, defined as "the study of the relationship between literature and the physical environment,"[13] allows us to understand how nature–society relationships are represented in texts, as well as opening up a discussion on representations of nature beyond texts.[14] Professor in philosophy and literature Ning Wang points out, the relationship between man and nature has long been a theme in Chinese literature.[15] It has therefore been unsurprising to see a growing trend of reading ancient and modern Chinese literature through an ecocritical perspective in the past decade.[16] The reasons why ecocriticism has become popular in China lie perhaps in the growing cross-disciplinary interactions within Chinese academia, as well as the increasing anxiety over environmental crisis that have been amplified by the media and discussed by Chinese intellectuals.[17] Meanwhile, the literature historian Qingqi Wei has underlined that reading contemporary Chinese literature through an ecocritical lens had already begun before the introduction of the concept of ecocriticism, as the environmental awareness has always been a theme in Chinese literature and philosophical thoughts, long before the modernization period.[18] In this context, rereading the works of "Obscure Poetry" from an ecocritical perspective is appropriate, not only to exemplify the ecological awareness in modern Chinese literature, but also to position Obscure Poetry in Chinese environmental discourses.

Three questions arise: first, in considering the fact that the poets in the "Obscure Poetry" group shared similar experiences of living in both rural and urban China, how do we understand the role of 'nature' in their poems? Second, how is 'nature' represented in relation to the theme of the 'self,' one of the major themes of "Obscure Poetry"? Finally, what are the possible influences that Obscure Poetry had on the development of Chinese environmental discourses since the 1970s?

With these questions in mind, I will offer an ecocritical reading of three of the most iconic poems/essays by two poets active in the "Obscure Poetry" group, Gu Cheng and Beidao.[19] These poems/essays were mostly published in *Today*, and were later edited into anthologies of "Obscure Poetry."[20] The popularity of these early anthologies in the 1980s was unprecedented. The editors for the second edition of "Selected poems of Obscure Poetry" stated in the afterword:

> The anthology was a special outcome in a special period. The strong and long-lasting impacts it had on the silent Chinese poetry perhaps exceeded all the other

> anthologies of poetry at the time. As a classic of poetry, the anthology is irreplaceable for its readers as it companied them for their journey of life . . . Even for those who no longer read poems today, the anthology remains in their memory.[21]

Although Gu Cheng and Beidao published their own volumes of poetry in later years, their poems are still mostly associated with "Obscure Poetry" and continue to appear in new editions of "Obscure Poetry" anthologies.[22]

The Generation

Li notes that the socio-cultural climate in China in the 1970s was unique for the generation which grew up in these ten years: they stepped into an historical interstice when they were barely teenagers and matured into adulthood in great social and political instability.[23] As aforementioned, the poets who later engaged in the underground poetry scene in Beijing had mostly gone through the "educated youth" program, which entailed displacing urban youths to the countryside where they learned about life and society through empirical lessons, hard labor and rural lifestyles. Attitudes towards this dislocation were mixed. While being driven away from the urban and political center made these youths' future appear uncertain, literature and other cultural/spiritual mediums fit in harmoniously with the rural life which brought them closer to their interior worlds and natural surroundings.[24]

Literature, and in particular poetry, thus became vehicles to contain and communicate such mixed, sometimes ambivalent feelings towards the poets' shifting situations and consequently their changing worldviews. One of the representative poets in "Obscure Poetry," Gu Cheng (1956–1993), spoke about the sense of living in a rural environment which led to the birth of "*A Fantasia to Life*," one of his most famous works:

> I left Beijing in 1969, herding pigs in Changyi County in Shangdong Province. There were very few people. Only large areas of deserted land with white saline-alkali soil, not even grass. I stood there between the sky and the land, feeling like there was nothing, but me. I knew I was small and insignificant, just like insects and grass. But my heart was not grass or stone. When the spring came, a feeling of life woke up and rose in my heart . . . The summer when I was turning fifteen, I wrote "A Fantasia to Life": *Without purpose/rippling in the blue sky/let the waterfall of sunshine/wash my skin brown . . . The sun toasts the earth/like toasting a piece of bread/I am walking/with barefoot/I left my footprints/marking the land like stamps/the world then permeates/my life*. When I became the world, the world became me as well . . . This is a good feeling of nature and life. This is the dream, light and flower derived from life, when one is young.[25]

Here a sense of loss can be detected, yet accompanied by a feeling of surrender. Yuko argues that during the time Gu Cheng and his family lived in the countryside, 'nature' replaced 'books' in Gu Cheng's spiritual world and inspired him to construct a 'natural ego.'[26] In Gu Cheng's prolific early production, 'nature' was used frequently and freely through personification, although at times it could be politically sensitive. For instance, in *A Fantasia to Life*, he wrote: "The sun is my tracker/it drags me/with its towline of light/step by step/completing the road trip in twelve hours/I am pushed by the wind/east and west/the sun disappeared in the dusk."[27] The ones who lived in China in that time would know that 'the sun' in literary language almost exclusively referred to one particular political leader. Had it been in Beijing or other big cities, the image Gu Cheng painted of 'the sun' would be deemed highly inadequate and could potentially bring disasters to the family.[28] However, being far away in the countryside, Gu Cheng was allowed to develop his language and imagination freely. As Gu reflected, he was lucky to be growing up in that era when a cultural vacuum was available in the middle of all the chaos.[29]

When Gu's family returned to Beijing in 1974, the poet's 'natural ego' was forced to change again. For the 'voice of nature' cannot communicate with people in the city any longer. He made great efforts to embrace the new surroundings, at the same time, to search for a new 'self': "I search for my own voice, but I also want to talk to people; this is a paradox."[30] The absence of 'nature' again left the poet in the darkness. This was the time he wrote the famous two-lines-poem, *A Generation*: "Dark night gave me dark eyes,/But I used them to look for light."[31] The "light" the poet was searching for was later described by himself as a 'cultural ego' that could be easily embraced and assimilated by a larger collective 'we,' and consequently a society that he and his companions can participate with revolutionary ideas.

If 'nature' in the rural world so harmoniously co-existed with the poet and was a source from which he drew inspiration, the attempt to use 'culture' as a channel to reach the outer society that Gu Cheng felt so alienated from was not successful. The unconventional linguistic formula he was using so freely was no longer communicable in the city and a deep anxiety forced him to look beyond both the 'natural ego' and the 'cultural ego.' In this process the 'ego' collapsed and led to a 'selfless' status.[32] Gu Cheng moved to New Zealand in 1987 and tried to reenact the lost 'nature' that was so significant for him in his teenage years. Yet he discovered more paradoxes than harmonies:

> I have a faith in nature . . . I feel I should not have more illusions when I am in nature and the natural beauty of my life would emerge there. Like the words in my poem: Wind is swinging its leaves/grass is growing its seeds/it is just nice when we are standing without words. This is an imagination. Nature is not nice. There are rats and flees in nature. It is not the 'nature' we see when we are on holiday. There is no electricity, no water, and no civilization. In this condition, you must battle with nature day and night.

Until one day he discovered that 'nature' is a state of mind:

> I woke up like a baby, because unconsciously I had forgotten about my thoughts and myself completely . . . when my heart calms down, I can see everything. The mountain is still mountain; the water is still water. Nothing has changed or disappeared. But I see them now.[33]

Gu Cheng believed that he reached the true "state of nature" when he finally abandoned all the purposes of the 'I.'[34] Gu Cheng reasoned his sense of belonging in 'nature' as a case of the 'Eastern' philosophy, in contrast to the 'Western' ideas of 'nature.'[35] Such exploration of relationships between one's inner self and 'nature' illustrates the extent to which young intellectuals in the 1970s started to participate in the circulation and translation of different representations of 'nature.' Similarly in Ah Cheng's *King of Trees*[36] published in 1986, the main characters experience individual emotional stories that are intertwined with clashing representations of 'nature.' In these literary works we can see a process of contemplating on and struggling between different ideas of 'nature' in the 1970s that is still present in today's discussion around human's position in relation to 'nature.' Noticeably, during this process the idea of 'self' was intertwined with the awareness of the question of 'nature,' which was reflected through other poems of "Obscure Poetry."

Nature's Child

Michelle Yeh observes that a prevalent and significant persona that Obscure Poetry poets used in their poems is that of the 'child.'[37] Through the viewpoint of a child, the poets were able to utter their 'new voice' and see 'new things' in the world. This position of a 'child' is constructed in relation to an adult. For adults, nature is just an arena of resources, whereas for children, nature is full of imaginaries and "offers a refuge for the oppressed and the weary."[38] Another representative poet Mang Ke writes in *The Moon on the Road* (1973): "The moon is walking me home/I want to bring her to the days in the future/it is quiet on the road . . . The moon is floating alone in the wild/ when did she run away/I had no idea." Here 'the moon' symbolizes a female, or even a motherly figure, that offers guardians and protection. Unable to locate the missing moon, the poet soon became aimless. The language is simple and direct, which is why Mang Ke was well known for his name as 'nature's child.'[39] Yeh notes, "Nature in Obscure Poetry is portrayed in a predominantly positive way and is often contrasted to negative social systems. Nature promises freedom from inhibitions imposed by society."[40] In these poems, one gets a sense that despite the inevitable fall of innocence in adulthood, there is still hope if one can remain a child in Mother Nature's arms.

I suggest that there is a connection between the poets' self-identity as children of 'nature,' and their desire to establish a new 'self' that sits in-between individualism and collectivism. This may not be so straightforward at the first sight. Unlike the previous political propaganda poetry, Obscure Poetry expressed a spirit of rebellion, as well as searched for a cultural heritage of independence in the ancient Chinese intellectuals (e.g. the ancient poet Qu Yuan (c. 340–278 BC).[41] Some of the poems were boldly expressing the opposition between 'I' and the world. The most famous example was Beidao's *The Answer*:

> I came into this world/Bringing only paper, rope, a shadow,/To proclaim before the judgment/The voice that has been judged:
>
> Let me tell you, world/I—do—not—believe!/If a thousand challengers lie beneath your feet,/Count me as number thousand and one.
>
> I don't believe the sky is blue/I don't believe in thunder's echoes/I don't believe that dreams are false/I don't believe that death has no revenge.
>
> (Beidao, *The Answer*, 1976, translated by Bonnie S. McDougall)

The dramatically different style was attacked for being 'incomprehensible' and too 'individualistic'; the name 'obscure' was given to the group.[42] However, as Ge argues, the 'self' expressed by Obscure Poets is not merely the individual self, but a self that is always part of a group.[43] Gu Cheng elaborated on this point and said:

> I think our poems are different because there is a 'self' that has contemporary young characters. The art and poetry we had in the past always promoted another kind of 'self'—a self-diminishing 'self'. (Through our poems) a new 'self' is born on the abandonment of this old 'self'. Only by loving oneself, and loving to become 'self', we can love everyone else, the nation, the life and the nature.[44]

Through the form of 'Nature's Child,' Obscure Poets found their own way of formulating a 'self' that was at once individual and collective. By situating oneself in relation to other beings, human or non-human, the Obscure Poets explored the themes of love, life and truth in a broader context. The interstitial positions between one and the other, human and nature, as well as individual and collective, began to be addressed implicitly. The still ambiguous self-identification, however, was further developed by poetry and literature groups after the "Obscure Poetry," which made more explicit connection with the ecological theme. This in turn, can be seen as contributing to the contemporary China's environmental discourse.

Obscure Poetry in Environmental Discourse

Ecological critical thinking and environmental consciousness have been present in Chinese literature since ancient times and this aspect has recently been added into Chinese ecocritical studies.[45] In the case of "Obscure Poetry," although 'nature' was not a strong theme, it was a symbolic thread that could shed some light on an often-neglected background in which different representations or understandings of 'nature' were produced. Reading these poems through an ecological or ecocritical lens enables us to reflect on the fact that, when the natural environment in 1970s China went through dramatic changes, these poems illustrated cultural and political dimensions of human–nature relationships. This is in contrast to the early ecological writings in the West (mainly US) in the 1960s and 1970s, where environmental crisis was the central focus, as exemplified by Rachel Carson's 1962 *Silent Spring*. One can then perhaps argue that there is a so-called Chinese ecocriticism that integrates both native and foreign ideas on the relationships between literature and nature.[46]

My contention here is that the concerns shown in the Chinese literature scene in 1970s about human–nature relationships can also be manifested in other forms than the direct concerns over environmental issues, as for instance, in the experience of young intellectuals developing their interstitial (individual–collective) identities. The poets which emerged after the "Obscure Poetry" in the mid-1980s categorized the intellectuals after the founding of the People's Republic of China in 1949 into three groups. The ones who were mostly writing political propaganda poems before 1970s were "the First generation," the "Obscure Poetry" poets belonged to "the Second Generation" and the ones after them were "the Newborn Generation" or "the Third Generation." Although the three groups' poetry works present very different features, and the latter two especially aimed to challenge earlier poetry, they did share the desire to nurture a sense of belonging. As Yeh recognizes, "the sometimes playful or carnivalesque side of the discourse cannot hide the intense desire for distinction through collective self-definition."[47]

However, such identity building shows us how the young intellectuals reacted to the socio-cultural and political changes in 1970s China, and in particular how this indirectly reflected on environmental changes. Kerridge argues that the aim of ecocriticism studies is that "environmental criteria will become an expected part of the debate about all kinds of new artistic work, and this will be a sign of a general shift in cultural values and—most importantly—in everyday behaviour."[48] Having this in mind, I suggest that we should also recognize the potential of all media and arts on shaping our ideas and representations of 'nature'; and by bringing such effects to light, we participate in the discourse of environment, thus influencing or potentially changing the way we engage with 'nature' in our daily life. The literary productions of "Obscure Poetry" offer such a contribution to environmental discourse, in that the poems became so influential and provocative that, almost all subsequent literature groups advanced or responded to their ideas, styles and themes.

Two literature groups in particular, "Scar literature" and "Root-seeking poetry," were established after "Obscure Poetry" in the 1980s. While "Scar literature" expresses "a longing for individual dignity and freedom . . . in an elegiac, condemnatory, or heart-wrenching tone,"[49] "Root-seeking poetry" emphasizes a reinterpretation of the roots of Chinese culture and aims to redefine Chinese identity.[50] Both can be seen as extensions of "Obscure Poetry," for they are also dedicated to discovering the essence of life, spirit and beauty, and reconnecting with a lost past.[51]

Yet groups emerged later that also challenge these laments of a lost 'nature' shown in the "Second Generation." For instance, the "Newborn Generation" took a different approach to responding to the "Obscure Poetry." In contrast to the idea that nature is a refuge from society, the poetry of "the Newborn Generation" shifts its focus to the urban area and expressed an "urban skeptic" or "frustrated urbanite."[52] In "Obscure Poetry" natural images are commonly used, whereas in "the Newborn Generation" there is no obvious empathy towards nature. Instead, a distance and separateness between human and nature is often portrayed in their work, to illustrate the harsh reality they were living in. This anti-romantic image is in fact also part of the contemporary environmental discourse in China, especially profound in the reality of modernization, urbanization and secularization in expanding city areas. Wang states that cities have gradually become our 'second nature' and the medium through which humans are indirectly interacting with the 'first nature.'[53] Both the 'first nature' and the 'second nature' are referring to a certain kind of representation of 'nature.' Moreover, individuals who are engaged with these different representations of nature may come across each other and exchange their ideas on 'nature,' which potentially leads to new meanings and perceptions. Contemporary Chinese literature is where we can trace such encounters and outcomes.[54]

Scholars in literary studies and ecocritical studies have generally overlooked the role of "Obscure Poetry" in forming and contributing to environmental discourses in China. While some criticize "Obscure Poetry" for not fitting into the social norms of the time and being therefore irrelevant to broader social changes, I argue the opposite.[55] The intellectuals who produced "Obscure Poetry" were in fact able to take advantage of their social-status (their well-off family background) and voice critical issues in society through their poems.[56] Therefore the works of "Obscure Poetry" indeed display a strong sense of responsibility and commitment towards society.

This chapter explored links between literature and 'nature' in contemporary China which have not been addressed in the ecocritical scholarship through a reading of the works of "Obscure Poetry." Incubated as an underground group in the later part of the Cultural Revolution, "Obscure Poetry" was established by a group of intellectuals during the period of the 1970s and early 1980s, and became known to the public by launching a privately published and disseminated journal *Today* in 1978. "Obscure Poetry" produced much poetry that would greatly influence Chinese literature in the post-Cultural Revolution era. By examining the socio-cultural and political background in which the group arose, I argue that the formation and production of "Obscure Poetry" can help us to understand the representations of nature in China in the historical interstice of the 1970s.

I use the recent "Chinese ecocriticism studies" to re-read the literary production of "Obscure Poetry," exemplified in the poems of Gu Cheng and Beidao; not only to bring out the 'nature' themes in the works of the group, but also to develop a style of ecocriticism that integrates Chinese literature aesthetics and its own ecocritical traditions.[57] In doing so, I suggest a different domain of possibility in viewing human's concerns of nature that differs from a focus on environmental crisis. In the works of "Obscure Poetry," we see a constant searching for a balanced relationship between the 'self' and the 'other,' an individualist 'I' and a collective 'We.' 'Nature' is represented in this search as a symbol for holistic forms of belonging, an environment that allows coexistence and co-creation. This representation of nature is far more than a simple continuation of the ancient Chinese philosophy of 'human and nature are one harmonious entity,' rather it is a longing for the freedom of expressing one's togetherness with one's surrounding, which goes beyond the boundaries of self/other, nature/culture and body/soul. In summary, the "Obscure Poetry" group and their works have encouraged later writers and poets to express their concerns about the pragmatic and destructive ideologies such as the developmentalist attitudes towards nature after the Open & Reform in 1978, as well as evoked other responses and reflections on a more nuanced picture of how 'nature' could be represented, such as in the works of the "Third Generation."

I believe more can be done in the future by using ecocriticism as a critical methodology in reading Chinese literature, especially the literature produced during the "educated youth" program. The transition from the urban to rural certainly played a curious role in inspiring the desire to express ideas artistically. Novellas, fictions and poems created in this period could be a fruitful platform to inquire how 'nature' was perceived and represented in this transition.

Notes

1 Tuo Li, "Editors' Words," in *The Seventies*, eds. Beidao and Tuo Li (Hong Kong: University of Oxford, 2008), 4.
2 Christopher Karlin Tong, "Beyond the Iron House: Ecological Thinking in Modern Chinese Literature, 1917–1937,"(University of California, Davis, 2014).
3 Jincai Yang, "Ecological Awareness in Contemporary Chinese Literature," *Neohelicon* 39 (2012), 107–118.
4 Shiqiang Wang, "Exploring the Emergence of 'Pre-Obscure Poetry' in the 1960s-1970s," *Yangzijiang Criticism* 2 (2014), 74–80.
5 Wang, "Exploring the Emergence of 'Pre-Obscure Poetry' in the 1960s-1970s."
6 Bonnie S. McDougall, and Kam Louie. *The Literature of China in the Twentieth Century* (London: Hurst & Company, 1997).
7 McDougall and Louie, *The Literature of China.*
8 Jeffrey Twitchell, and Fan Huang. "Avant-Garde Poetry in China: The Nanjing Scene 1981–1992," *World Literature Today* 71, no. 1 (1997), 29–35.
9 McDougall and Louie, *The Literature of China.*
10 Twitchell and Huang, "Avant-Garde Poetry in China," 29.
11 Beidao. "Out of the Cradle, Endlessly Sleepwalking," *World Literature Today* 82, no. 6 (2008), 22.

12 McDougall and Louie, *The Literature of China.*
13 Cheryll Glotfelty, and Harold Fromm. *The Ecocriticism Reader: Landmarks in Literary Ecology* (Athens and London: University of Georgia Press, 1996), xix.
14 Hannes Bergthaller, "The Canon of East Asian Ecocriticism and the Duplicity of Culture," [In English]. *CLCWeb: Comparative Literature and Culture* 16 (2014). Christopher Cokinos, "What Is Ecocriticism? In Defining Ecocritical Theory and Practice: Sixteen Position Papers from the 1994 Western Literature Association Meeting." <www.asle.org/wp-content/uploads/ASLE_Primer_DefiningEcocrit.pdf>
15 Ning Wang, "Toward a Literary Environmental Ethics: A Reflection on Eco-Criticism," *Neohelicon* 36, no. 2 (2009), 289–298.
16 Ning Wang, "Global in the Local: Ecocriticism in China," *ISLE: Interdisciplinary Studies in Literature and Environment* 21, no. 4 (2015), 739–748.
17 Douglas Scott Berman, "Chinese Ecocriticism: A Survey of the Landscape," *Literature Compass* 12, no. 8 (2015), 396–403.
18 Qingqi Wei, "Chinese Ecocriticism in the Last Ten Years," in *The Oxford Handbook of Ecocriticism*, ed. Greg Garrard (New York: Oxford University Press, 2014), 537–546.
19 These two poets were chosen for their names are among the most well-known poets in mainland China, and their poems are also the most accessible ones among other "Obscure Poetry" poets' works.
20 The first anthology that uses the name "Obscure Poetry", *Selected Poems of Obscure Poetry [Meng long shi xuan]* was edited by Yuejun Yan, Yan Gao, Yun Liang and Fang Gu, and published in November 1985 by Chunfeng Wenyi Press. Another early edition of "Obscure Poetry" is Daxiang Yu, and Qiuling Liu (eds.). *Selection of Obscure Poetry [Meng Long Shi Jing Xuan]* (Huazhong Normal University Press 1986). It is important to note here that there were female poets in the group such as Shu Ting that also were publicly renown.
21 Yan et al. *Selected Poems of Obscure Poetry*, second edition, 2002.
22 See Guangwei Cheng, and Zicheng Hong, eds. *New Edition of Obscure Poetry [Meng Long Shi Xin Bian]* (Wuhan: Changjiang Wenyi Press, 2004).
23 Tuo Li, "Editors' Words." in *The Seventies.*
24 Shiqiang Wang, "Exploring the Emergence of 'Pre-Obscure Poetry' in the 1960s-1970s," *Yangzijiang Criticism* 2 (2014), 74–80.
25 Cheng Gu, "From Self to Nature—Lecture One," in *Selected Essays of Gu Cheng* (Tianjin: Baihua Wenyi Press, 1993), 264.
26 Shima Yuko, "On Gu Cheng's Ego and His Language of Poetry," *Jianghan Academic* 33, no. 2 (2014), 69–77.
27 The poem was written in 1971, when Gu Cheng was only fifteen.
28 Gong Gu, "Two Generations—Starting From 'Not Understanding' of Poetry," *Poetry* 10 (1980), 49.
29 Gu, "From Self to Nature."
30 Ibid.
31 This iconic poem was written in 1979.
32 Gu, "From Self to Nature."
33 Ibid.
34 Cheng Gu, *The 'I' Without Purpose—Themes of a Natural Philosophy*. Selected Essays of Gu Cheng Four (Beijing: China Culture Press, 2007).
35 Gu, *The 'I' Without Purpose.*
36 Ah Cheng. *The King of Trees*. Translated by Bonnie S. McDougall (New York: New Directions Books, 2010. 1984).
37 Michelle Yeh, "Nature's Child and the Frustrated Urbanite: Expressions of the Self in Contemporary Chinese Poetry," *World Literature Today* 65, no. 3 (1991), 405–409.
38 Beidao's poem, cited in Bonnie S. McDougall, "Beidao's Poetry: Revelation and Communication," *Modern Chinese Literature* 1, no. 2 (1985), 225–252.
39 Pingqiao Lin, "A Naked Appearance on 'Son of Nature': A Brief Comment on Mang Ke's Undergroud Poem," *Journal of Beijing University of Technology* 8, no. 6 (2008), 68–72.
40 Yeh, "Nature's Child," 407.
41 Mao Nie, "Cultural Heritage and Spirtual Rebellion of the Obscure Poetry," *Journal of Center South University* 22, no. 1 (2016), 190–196.

42 Ming Zhang, "The Annoying 'Obscurity'," *Poetry*, no. 8 (1980), 8.
43 Naifu Ge, "On Obscure Poetry," *Journal of Hainan Normal University* 15, no. 1 (2002), 50–55.
44 Cheng Gu, "Please Listen to Our Voices," in *Young Poets on Poems*, ed. Laomu (Beijing: Beijing University Wusi Literature Association, 1985).
45 Zhiwen Deng, "Historical Consciousness Regarding the Construction of Chinese Ecocriticism," *Humanities & Social Science Journal of Hainan University* 31, no. 2 (2013), 30–36.
46 Qingqi Wei, "Chinese Ecocriticism in the Last Ten Years," in *The Oxford Handbook of Ecocriticism*, ed. Greg Garrard, 539.
47 Michelle Yeh, "Anxiety & Liberation: Notes on the Recent Chinese Poetry Scene," *World Literature Today* 81, no. 4 (2007), 28–35, 32.
48 Richard Kerridge. "Ecocritical Approaches to Literary Form and Genre: Urgency, Depth, Provisionality, Temporality," in *The Oxford Handbook of Ecocriticism*, ed. Greg Garrard, 361.
49 Michelle Yeh. "Monologue of a Stormy Soul': The Poetry and Poetics of Duo, 1972–1988," *World Literature Today* 85, no. 2 (2011), 51–57, 52.
50 Mao Nie, "Cultural Heritage and Spirtual Rebellion of the Obscure Poetry."
51 Yeh, "Nature's Child."
52 Yeh, "Nature's Child."
53 Wang, *A Study of Natural Spirit.*
54 Wang, *A Study of Natural Spirit.*
55 Wei Su, and Wendy Larson. "The Disintegration of the Poetic Berlin Wall," in *Urban Space in Contemporary China: The Potential for Autonomy and Community in Post-Mao China*, eds. Deborah S. Davis, Richard Kraus, Barry Naughton, and Elizabeth Perry (New York: Cambridge University Press, 1995), 279–294.
56 Wang, *A Study of Natural Spirit.*
57 Ming Zhang, "Super Circulation Development of Chinese Eco-Criticism," *Journal of Shaoyang University* 11, no. 2 (2012), 90–95.

References

Cheng, Ah. *The King of Trees*, Bonnie S. McDougall (trans.). New York: New Directions Books, 2010. (1984).

Dao, Bei. "Out of the Cradle, Endlessly Sleepwalking." *World Literature Today* 82, no. 6 (2008): 22.

Berman, Douglas Scott. "Chinese Ecocriticism: A Survey of the Landscape." *Literature Compass* 12, no. 8 (2015): 396–403.

Cokinos, Christopher. "What Is Ecocriticism? In Defining Ecocritical Theory and Practice: Sixteen Position Papers from the 1994 Western Literature Association Meeting." retrieved from www.asle.org/wp-content/uploads/ASLE_Primer_DefiningEcocrit.pdf.

Deng, Zhiwen. "Historical Consciousness Regarding the Construction of Chinese Ecocriticism." *Humanities & Social Science Journal of Hainan University* 31, no. 2 (2013): 30–36.

Garrard, Greg (ed.). *The Oxford Handbook of Ecocriticism*. New York: Oxford University Press, 2014.

Garrard, Greg. *Ecocriticism*. London: Routledge, 2012.

Ge, Naifu. "On Obscure Poetry." *Journal of Hainan Normal University* 15, no. 1 (2002): 50–55.

Glotfelty, Cheryll, and Harold Fromm. *The Ecocriticism Reader: Landmarks in Literary Ecology*. Athens and London: University of Georgia Press, 1996.

Gu, Cheng. "From Self to Nature—Lecture One." in *Selected Essays of Gu Cheng*, 264. Tianjin: Baihua Wenyi Press, 1993.

Gu, Cheng. *The 'I' Without Purpose—Themes of a Natural Philosophy*. Selected Essays of Gu Cheng Four. Beijing: China Culture Press, 2007.

Gu, Cheng. "Please Listen to Our Voices." in Laomu (ed.), *Young Poets on Poems*. Beijing: Beijing University Wusi Literature Association, 1985.

Gu, Gong. "Two Generations—Starting From 'Not Understanding' of Poetry." *Poetry* 10 (1980): 49.

Kerridge, Richard. "Ecocritical Approaches to Literary Form and Genre: Urgency, Depth, Provisionality, Temporality." in Greg Garrard (ed.), *The Oxford Handbook of Ecocriticism*, 361–376. New York: Oxford University Press, 2014.

Li, Tuo. "Editors' Words." in Beidao and Tuo Li (eds.), *The Seventies*, 1. Hong Kong: University of Oxford, 2008.

Lin, Pingqiao. "A Naked Appearance on 'Son of Nature': A Brief Comment on Mangke's Undergroud Poem." *Journal of Beijing University of Technology* 8, no. 6 (2008): 68–72.

McDougall, Bonnie S. "Beidao's Poetry: Revelation and Communication." *Modern Chinese LIterature* 1, no. 2 (1985): 225–252.

McDougall, Bonnie S., and Kam Louie. *The Literature of China in the Twentieth Century*. London: Hurst&Company, 1997.

Nie, Mao. "Cultural Heritage and Spirtual Rebellion of the Obscure Poetry." *Journal of Center South University* 22, no. 1 (2016): 190–196.

Tong, Christopher Karlin. *Beyong the Iron House: Ecological Thinking in Modern Chinese Literature, 1917–1937*. Berkeley: University of California, Davis, 2014.

Twitchell, Jeffrey, and Fan Huang. "Avant-Garde Poetry in China: The Nanjing Scene 1981–1992." *World Literature Today* 71, no. 1 (1997): 29–35.

Wang, Ning. "Toward a Literary Environmental Ethics: A Reflection on Eco-Criticism." *Neohelicon* 36, no. 2 (2009): 289–298.

Wang, Shiqiang. "Exploring the Emergence of 'Pre-Obscure Poetry' in the 1960s-1970s." *Yangzijiang Criticism* 2 (2014): 74–80.

Wang, Shudong. *A Study of Natural Spirit in Contemporary Chinese Literature*. Haerbin: Heilongjiang People Press, 2004.

Wei, Qingqi. "Chinese Ecocriticism in the Last Ten Years." in Greg Garrard (ed.), *The Oxford Handbook of Ecocriticism*, 537–546. New York: Oxford University Press, 2014.

Wei, Su, and Wendy Larson. "The Disintegration of the Poetic 'Berlin Wall'." in Deborah S. Davis, Richard Kraus, Barry Naughton and Elizabeth Perry (eds.), *Urban Space in Contemporary China: The Potential for Autonomy and Community in Post-Mao China*, 279–294. New York: Cambridge University Press, 1995.

Williams, Raymond. *Problems in Materialism and Culture: Selected Essays*. London: Verso, 1980.

Yang, Jincai. "Ecological Awareness in Contemporary Chinese Literature." *Neohelicon* 39 (2012): 107–118.

Yeh, Michelle. "Anxiety & Liberation: Notes on the Recent Chinese Poetry Scene." *World Literature Today* 81, no. 4 (2007): 28–35.

Yeh, Michelle. "Monologue of a Stormy Soul': The Poetry and Poetics of Duo, 1972–1988." *World Literature Today* 85, no. 2 (2011): 51–57.

Yeh, Michelle. "Nature's Child and the Frustrated Urbanite: Expressions of the Self in Contemporary Chinese Poetry." *World Literature Today* 65, no. 3 (1991): 405–409.

Yuko, Shima. "On Gu Cheng's Ego and His Language of Poetry." *Jianghan Academic* 33, no. 2 (2014): 69–77.

Zhang, Ming. "The Annoying 'Obscurity'." *Poetry* no. 8 (1980): 8.

Zhang, Ming. "Super Circulation Development of Chinese Eco-Criticism." *Journal of Shaoyang University* 11, no. 2 (2012): 90–95.

12 The Green Corridor

A Vision for Lisbon

Maria João Matos

Aiming to understand the relevance of landscape design strategies and concepts in trans-industrial urban regions, this chapter explores the evolution of the place of nature in the urbanized territory of Lisbon. It will highlight the upsurge in environmental awareness and the emergence of a new relation between city and nature as exemplified by a central green corridor. *Parkways* and *greenways* are essential to this discussion; within the framework of the field of landscape design, they function as operative tools for urban and territorial planning. In origin, the American parkway was defined as a scenic road set out in a preserved forest or a landscape corridor,[1] designed exclusively for pedestrian, carriages and horseback use.[2] As the notion of the parkway gradually changed to refer more generally to 'a road within a park,'[3] the notion of greenway as a nature-related and car-free space gained importance, following innovations by Frederick Law Olmsted,[4] a key figure in landscape design.[5] The landscape protection specialist Charles E. Little and the landscape architect Jack Ahern[6] present the greenway as a complex notion, which evolved as an element of metropolitan systems at the end of the 19th and the beginning of the 20th centuries, when landscape was often a central instrument in urban planning. In the 1960s, according to Ahern,[7] the concept of greenway gained a particular relevance, related to its use in ecological based planning. In the following decade, the introduction of the concept of the "greenline" as "a new idea in land protection and management based on mixed public—private ownership," that evolved from "large, park-like reserves to linear corridors including historic canals, railroads, and rivers,"[8] coincided with a new urban development stage, characterized by a decrease in public funding, combined with the urge to preserve nature within metropolitan areas.[9] The same author situates the focus on greenway systems, in the American metropolitan areas of the 1980s, as a new strategy aiming to respond to the loss of open-air spaces and to fulfill the need for recreation within urban areas.[10]

Lisbon's urban planning, throughout the last two centuries, presents the same phases, in regard to the changing roles of greenway systems. Recently, with the rise of Lisbon as a trans-industrial *metapolis*,[11] Gonçalo Ribeiro Telles[12] became the driving force behind a new approach to planning in which particular landscape design instruments and concepts—such as the Green Plan and the Green Corridor—took a decisive role for the city's sustainable development. This chapter therefore presents the evolution of Lisbon's green spaces, focusing mainly on the changes that occurred to green structures in the trans-industrial socioeconomic context of the 1970s, when existing interstitial spaces progressively assumed the form of green corridors, playing a significant role in the greening of the urban fabric.

From Urban Parks to Trans-industrial Landscapes

The first public urban parks appeared in the 19th century in the United Kingdom, using natural models in order to provide cities' inhabitants with contact with nature. They were also designed as a response to issues related to uncontrolled urban growth, which had exploded during the Industrial Revolution. From that moment on, urban parks multiplied and evolved in their form and uses, helping to reshape cities.[13] European parks were a main inspiration for the American landscape designer F. L. Olmsted,[14] whose "vision almost alone led the American nation from the concept of the isolated urban park to that of city and country as being a single design."[15] He introduced typologies such as the parkway, a concept he and his partner Charles Eliot experimented with in Boston's "Emerald Necklace," a "model of integration of existing protected lands, ecological corridors, and built linear elements,"[16] which we can relate to contemporary notions of sustainable development and resilience, applied to urban and territorial management.

In the aftermath of the Second World War, most of the Western world was living under democratic political systems, combined with economies based on the development of mass production industries and consumption. Building new urban areas became a major concern for cities with fast growing populations. The *Athens Charter*[17] had become a reference for modern cities' planning, offering a new model for a functional relation between urban and natural environments. It advocated the creation of wide public green spaces next to residential buildings, with the belief that agglomerations should become green cities with extensive green areas for healthy and recreational activities. The *Bos Park* in Amsterdam, which inspired the layout of the Monsanto Park in Lisbon, was, in some ways, a result of this modern vision. Furthermore, the concepts of "garden city" and "greenbelt" became important instruments in planning, used to control urban sprawl. By that time, in Paris, policies for the reinforcement of the *Ceinture Verte* (greenbelt) clearly reflected this tendency.[18]

With the acceleration of globalization processes in the 1960s and 1970s, cities and urban regions witnessed major changes in their structure and growth, accompanied by a technological and informational revolution[19] as well as an exponential increase in populations. As a result, intensified urban sprawl and natural resource destruction, combined with rejected open-air spaces originated by modern urban planning, led to a shift in the perception of the relation between urban and natural environments. At this socioeconomic tipping point, designated as trans-industriality, when industrialization subsists with resistance to it and the territory changes in ways that we cannot understand, the citizens' growing demand for green becomes evident, as the increased environmental awareness favors the development of an approach to territory founded on ecological principles and on the use of landscape design for territorial planning.[20] The establishment of the European Conservation year, in 1970, alerting to the finite capacities of the ecosystem, was a milestone in this paradigm shift,[21] reinforced by other events such as the publishing of *The Limits to Growth* by the Club of Rome[22] and the economic impact of the 1973 oil crisis. Furthermore, the notion of landscape was becoming more relevant when referring to urban identities, as a result of concerns about rapid territorial transformations and the consequent loss of the *genius loci*, the spirit of place.

The understanding of "the city" and its relation to nature started to change at the outset of the trans-industrial era. The German urban planner Thomas Sieverts,[23] the

French urban sociologist François Ascher[24] and others became aware of significant transformations taking place in metropolitan territories: they argued that it made no sense anymore to talk about cities in their traditional form, since the spaces where we were living were much more complex and diversified than the dense, centralized and continuous city that prevailed throughout the Industrial era. In this heterogeneous reality, new types of spaces would be combined with classical ones, establishing unusual morphological, economic, social and ecological relations among them. Emerging types of territories brought about new concepts—the *Zwischenstadt*,[25] the *metapolis*[26] (as previously described), the *palimpsest*,[27] the *terrains vagues*,[28] among others—novel ways to understand the urbanized territory and tools to reconfigure it. Another kind of relation with nature would be inevitable.

By that time, Olmsted's ideas were being rediscovered and adapted across the globe to provide beautiful sceneries, a close and balanced relation between cities and green spaces, as well as a way to solve environmental and functional issues in metropolitan territories.[29] Urban policies in 1980s Paris represent a remarkable case of success in the assembly of existing urban fragments, a major role being played by landscape design: different kinds of urban green spaces would restructure the metropolis and rehabilitate in-between spaces, as well as abandoned structures inherited from the industrial era. The recovery of the greenbelt,[30] the new *Parc de La Villette* and the green corridor called the *Promenade Plantée*[31] are only a few of the many spaces that came to reconfigure a new metropolis, reinterpreting Olmsted's idea of a structuring continuous green urban network and giving continuity to Forestier's vision for Paris.[32] Lisbon's urban policies on green spaces were to follow this tendency.

Lisbon and Nature: From the Industrial City to the Trans-industrial Metropolis

Until the end of the 18th century, the growth of Lisbon essentially took place along its waterfront. From then on, successive expansion projects, moving away from the riverside, were based on the idea of a green promenade, starting with the *Passeio Público*. Planned as early as 1764, to extend the central area of the city,[33] the *Passeio Público* became the first public urban green space in Portugal. It mirrored the trend of "public promenades" existing in other European cities.[34] Initially rejected by the public, it eventually became a symbol of significant changes in Portuguese society and mentalities in the mid-19th century.[35]

By then, the French architect Pierre-Joseph Pézérat published his *Mémoire sur les études d'améliorations et embellissements de Lisbonne*, inspired by Haussmann's reconfiguration of Paris, influencing new proposals for the expansion of the city.[36] One of these proposals was the extension of the *Passeio Público* further to the Northwest, laid out as a Haussmannian tree-lined boulevard: the *Avenida da Liberdade*.[37] Another was the implementation of the Eduardo VII Park, inspired by the Parisian informal gardens, for the north-western completion of this boulevard.[38]

From the very beginning, these extension proposals corresponded to the notion of greenway, in the sense that they were strips of green articulating the transition between the growing urban fabric and the still existing rural space in the immediate surroundings of the city. One can argue that these successive landscape projects, besides being inspired by contemporaneous European urban works, envisioned a green system for Lisbon, understood as a structuring basis for the expansion of the metropolitan area,

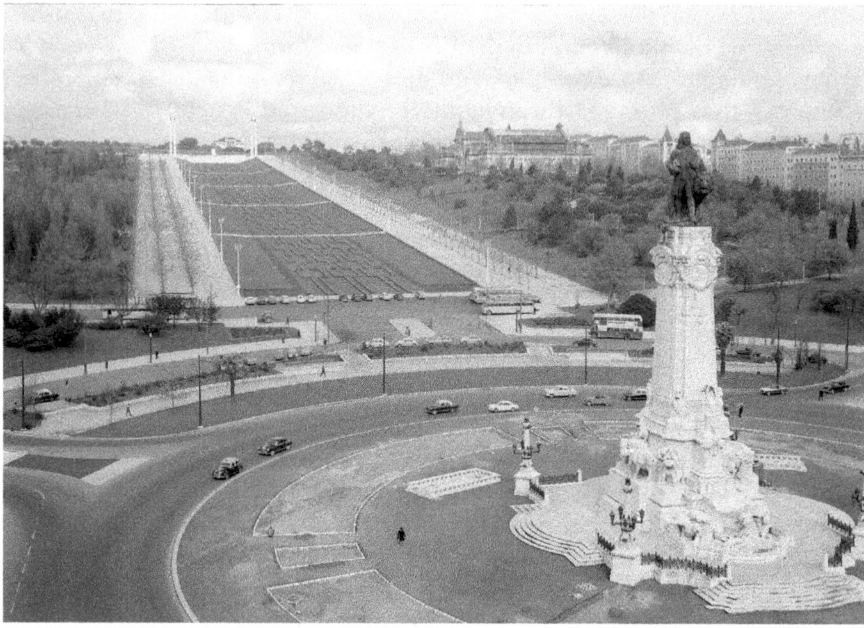

Figure 12.1 View of Eduardo VII Park from the City Center

Authorship: Horácio Novais Studio. Produced during the activity of the Studio (1930–1980). Source: Calouste Gulbenkian Foundation – Art Library

comparable to Olmsted's system of greenways and parks for Boston, despite the obvious differences in terms of scale, geography and socio-cultural context.

The creation of the Forest Park of Monsanto, even further to the Northwest of the consolidated city, was to be a major part of this system. Its seed was planted in the late 1860s by a report underlining the importance of this green lung for the city.[39] Two projects were later proposed: one by the Mac Bride brothers, two medical doctors who, in the 1920s, suggested a forest park at the culmination of a greenway, based on aesthetic, social and hygienist principles;[40] the other one by Jean-Claude Nicolas Forestier, a landscape engineer and author of *Grandes villes et systèmes de parcs*, who had proposed a system of parks for Paris. Greatly influenced by Olmsted's notions on park systems,[41] he designed a plan for Lisbon's extension consisting of a straight-line greenway, giving access to the "great park."[42]

The final step in the implementation of the forest park in Lisbon came with the dictatorial regime in 1926,[43] which intended to create a new image for the capital. In a context of aggressive urban policies and considerable investment in public works,[44] a law from 1934 finally established the creation of the Forest Park and opened the way for the necessary expropriations of land on the Monsanto Mountain, an arid hill of 1,000 hectares. The proposed green area, peripheral to the city, was supposed to become a strategic infrastructure for life quality, as well as for the reinforcement of the representative roles of the state and the capital city.[45]

The layout of the Forest Park of Monsanto, conceived by Keil do Amaral, was inspired by the European parks and gardens he visited in Paris, London and several Dutch cities, in particular by the modern and innovative *Bos Park* in Amsterdam, regarded as a pioneering example of a modern park for active recreation.[46] *Bos Park* aimed to provide a free experience of nature while involving citizens in the design process; it was environment oriented and based on sociologic studies on the uses of space.[47] Obviously, the Portuguese political and socio-cultural context was quite different from the Dutch one. Therefore, despite Keil's forward thinking, the process for the implementation of the park in Lisbon—intended for the recreation of all social classes—was marked by improvisation and the final result remained limited by a narrow vision and lack of technical knowhow.[48] The green structure—a forest crossed by traffic roads, punctuated by belvederes and modern sport and leisure facilities—was cut through by a new highway, linking the city center to the western limit of the metropolis, while pedestrian connections to the city were not a priority—an option that would later reinforce the split between the park and the urbanized area.

In the post-war period, the Portuguese Regime invested in industrialization. Migration to the capital city increased, along with metropolization processes,[49] causing the fast expansion of suburban areas from the 1950s onward.[50] In the next decade, the Colonial War in Africa absorbed resources that could not be invested in infrastructures, facilities or social housing, impacting life conditions.[51] As migration to Lisbon's Region intensified, a series of processes, such as uncontrolled urban sprawl

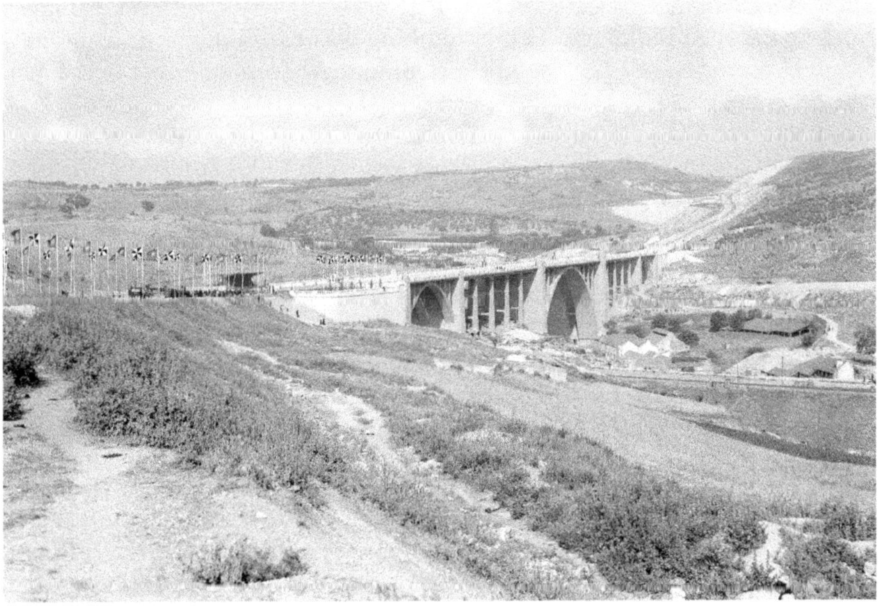

Figure 12.2 Monsanto and Highway, Linking the City Center to the Western Limit of the Metropolis

Authorship: Estúdio Horácio Novais. Most probably produced in the 1940s.

Source: Calouste Gulbenkian Foundation – Art Library

and illegally built neighborhoods, emerged, with damaging consequences for the territory.[52] These would only be clearly perceived and dealt with after Portugal entered the EEC.[53]

Earlier, Lisbon's 1948 Master Plan—also known as the De Gröer Plan—had proposed a greenbelt around the city, foreseeing the expansion of the Park of Monsanto. The idea never came into effect and the consequences of urban pressure became a real threat to Monsanto throughout the next decades, as legal and illegal construction progressively invaded the green area.

Measures for the preservation of nature and landscape only started being applied to the Park of Monsanto in 1974, thanks to a new law protecting the area in order to stop its decay. It initiated a slow recovery of its symbolic role, providing new uses for citizens in a park that could finally become what it was planned to be by Keil: a green space accessible to all, essential to the environmental balance of the Lisbon's region. Ultimately the Portuguese political and cultural revolution of 1974 enabled the capital city to embrace a trans-industrial socioeconomic paradigm, progressively facilitating in its management a reflection on the necessary interconnection between nature and city. This vision was to be reinforced by a new Master Plan, published in 1977, and by the definition of the limits of the park in 1979.[54] In the 1980s, the expansion of the metropolitan area slowed down, as the population got older and, at the same time, new lifestyles emerged.[55] Even though political and legal measures to protect the Forest Park were being taken, during this decade, Monsanto was hampered by the consequences of the previous uncontrolled urban expansion. The park became the geographical center of the metropolis and was severely damaged by construction, remaining a rejected part of the sprawling city,[56] with several interstitial areas and abandoned buildings. This green lung was essentially seen as a dangerous and decadent place, as well as a useful shortcut to drive from one area of the metropolis to the other.

The resistance to accept the decaying Forest Park as part of the city can be explained by the words of Sieverts, describing the *Zwischenstadt*: "This landscape where we live now, which is neither city nor country but has characteristics of both, has no suitable name nor is it visually remarkable."[57] Therefore, it took us some time to get used and pay attention to it. A *metapolis*, as defined by Ascher,[58] with its heterogeneous spaces and interstices, was taking shape in Lisbon's region. At the global level, territorial planning policies were rediscovering the importance of landscape as a key instrument for planning,[59] a phenomenon that started to show up in Lisbon's urban plans under the influence of Telles.

Bringing Nature to the City: The Green Corridor

As an outcome of the aforementioned ecological awareness, the 1990s witnessed the rise of urban sustainability concerns. In Europe, they gained considerable importance with the *Green Paper on Urban Environment* that stood for the implementation of environmental objectives in urban areas.[60] By that time, after Portugal had joined the EEC, a radical shift occurred in urban planning policies for Lisbon, which focused increasingly on urban competitiveness, while simultaneously expanding green structures and reinforcing the operative role of landscape as a support for sustainable development. One of the objectives of the 1994 Master Plan was to improve the city's environmental quality.[61] It proposed the connection of the major green areas of the city through greenways, defined as an environmental component of the Master Plan by Telles in Lisbon's Green Plan.[62]

Ahern refers to the global strategy for the implementation of Lisbon's Metropolitan Area greenway system as hinged on "a broad and inclusive understanding of greenway resources, both natural and cultural," articulating a continuum in which the combination of natural and cultural elements characterize a particular cultural landscape.[63] As part of this metropolitan vision, the Green Plan had its principles oriented towards the enhancement of history, "the universal sense and the Mediterranean expression of the city, the richness of its outskirts, the values of natural landscape and the highlight of the 'genius' and remarkable and unique nature of the site."[64] It was developed from three basic concepts: the *continuum naturale*, understood as a continuous system based on ecological principles and on the utility of landscape, that supports wildlife; the *cultural continuity*, integrating built and open spaces; the *genius loci*, including geology, topography, climate, the flow of water and their relation with the built environment.[65] Taking into account the segregated and fragmented situation of the Forest Park, Telles recommended its integration into the recreational spaces of Lisbon, suggesting a series of pedestrian connections with different parts of the city.[66]

The Green Corridor,[67] a greenway linking the central city to Monsanto, would be one of these connections.[68] A green urban system for Lisbon had actually been envisioned by Telles in the 1970s,[69] in tune with what was happening in other countries and aligned with a new attitude towards the role of landscape for urban planning. By then, the Green Corridor was framed as a landscape architecture project connecting *Avenida da Liberdade* with the Forest Park of Monsanto.[70] Telles would have to wait two decades to design his cherished Green Corridor as part of a broader plan, which formalized his reinterpretation of the concept created and applied by Olmsted: a green system where parks would no longer have strict limits, green spaces would be connected and these connections, or greenways, would be designed based on their integration within landscape.[71] For Telles, urban planning could not be performed in the traditional way anymore.[72] He believed that a new concept of urban design and green space was needed to embrace nature, within the city, as well as to connect the city with the surrounding landscape, which should be multifunctional, incorporating spaces of production, protection, culture and recreation. In the set out of green spaces, natural features, history and cultural references should be combined to ensure the economic viability and ecological sustainability of the urbanized territory, never forgetting to actively involve citizens in the conception, use and management of the places.[73] He coherently used these principles when he started to conceive the Green Corridor: a continuous structure that runs across the urban fabric, rooted in ecological and cultural principles, intended for the recreation of citizens, occupying *terrains vagues* and including zones with different ecological and geographical profiles.[74] This new type of greenway, although inspired by Olmsted's approach, reflects the changing role of nature and adapts to the interstitial spaces of the trans-industrial territories described, among others, by Ascher and Sieverts.[75] Ahern recognizes that "Telles' work anticipated the greenway concept, and identified the importance of green corridors, where resources are concentrated, to link natural and cultural landscapes in the region, including the city of Lisbon."[76]

The Green Corridor only started to take shape in the 1990s, with the conclusion of two separate green islands within the urban fabric: a naturalized space composed of autochtonous vegetation, laid out as a belvedere over the central city, and another

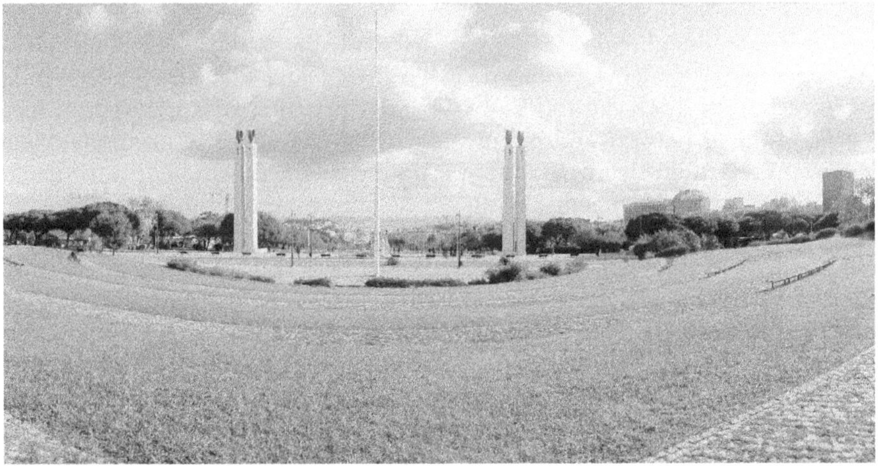

Figure 12.3 Amália Rodrigues Gardens – one of the green islands, completed in the 1990s as part of the Green Corridor, looking over Eduardo VII Park and the city center

Authorship: João Pagaime. 2016.

one containing allotments for agriculture.[77] The corridor was jeopardized by the approval of new construction, which would strangulate the greenway. Still, it was redesigned by the Municipality, although in some parts it had to be reduced, literally, to a "corridor" between buildings.[78]

In 2012, the revision of the Master Plan once more reinforced the integration of nature within the urban fabric, coinciding with the Green Corridor's completion, which played an essential role in bringing the city and the Forest Park closer to each other, both physically and conceptually. Moreover, in line with the growing importance of Monsanto as an integrated part of the city, as well as to respond to the demand for additional leisure spaces, new ones were inaugurated and several improvements were carried out within the Park.[79] As a consequence of these actions, combined with the change in the perceptions of citizens towards Monsanto, it is currently an area intensively used for sports, walks, cultural events and other leisure activities, a tendency that continues to grow. Telles' vision, inspired by Olmsted's approach, based on natural and built heritage preservation, as well as on the knowledge of natural systems and places, was a central piece for the integration of nature within Lisbon's *metapolis*, enabling its sustainable development.

Conclusion: Greenways for Sustainable Urbanized Territories?

> By establishing 'fingers of green' in the urban and suburban areas where people live, a physical connection is made which supports this philosophical or spiritual need for human—nature contact. When the human–nature access links with other resources, the benefits are multiplied in a synergistic manner, at least in terms of space utilisation.[80]

Today, Olmsted's notions remain relevant for urban and territorial planning, being readapted by leading figures in these fields, such as Telles or Desvigne.[81] The latter

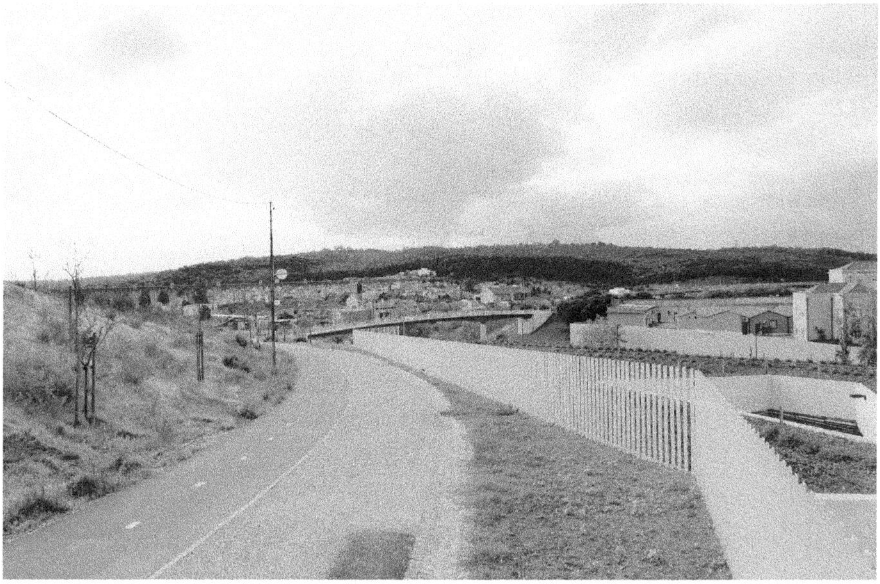

Figure 12.4 Recently Completed Part of the Green Corridor. Monsanto Forest Park in the background

Photograph by the author. 2016.

states that the American systems of parks of the 19th century, sometimes used to structure cities' growth, are also effective for structuring the European outskirts and their typology can now be transposed to reorganize urban sprawl, incorporating geography, infrastructures and industrial sites as references. He adopts Olmsted's approach to "site," rooted in the deep understanding of geography and natural elements, in line with pragmatic naturalists. His recent interventions recover this type of naturalist approach, which he believes to be compliant with the current perception of nature—clearly expressed in his project for the Cluster Paris-Saclay, a large-scale landscape project inspired by the Metropolitan Park System of Boston.[82]

As in Olmsted's times, landscape is often used as a central instrument for urban planning, responding to current ecological and quality of life concerns, as well as fulfilling the demand for biodiversity and natural beauty in urbanized contexts. Simultaneously, in contemporary cities, shaped by trans-industrial cultural and economic activities and inhabited by heterogeneous groups and individuals, landscape can serve as a vehicle for the construction of collective identities and social links. Therefore, urban renewal through landscape became a way for the political sphere to work together with different groups and individuals.[83] The fundamental principles of the *European Landscape Convention* openly support this objective.[84]

On the other hand, Ahern's statement cited above reminds us of the relevance of the greenway concept as a new strategy for the changing attitude towards the relation between city and nature, when large parks by themselves can no longer meet the need of nature within the urbanized context.[85] Also Batlle[86] observes that we are in need of types of exterior spaces capable to combine the old ideas of the systems of parks with the continuity of green ecologic systems at different scales—a strategy adopted by

Telles for Lisbon—essentially based on greenways. The use of this typology can therefore play an important role in associating landscapes with urban identity, enabling the inhabitants of trans-industrial cities to experience nature and helping to achieve the sustainable development of urbanized territories.

Notes

1 Oxford Reference. "Parkway," *Oxford Reference*. Oxford University Press. Accessed May 29, 2016. <www.oxfordreference.com>.
2 Charles E. Little, *Greenways for America* (Baltimore/London: The Johns Hopkins University Press, 1995), 12.
3 Geoffrey Jellicoe and Susan Jellicoe, *The Landscape of Man. Shaping the Environment from Prehistory to the Present Day* (London: Thames and Hudson, 1995), 311.
4 F. L. Olmsted conceived several remarkable parks, among others, the Central Park in New York (1858–1876) and the Prospect Park in Brooklyn (1865–1895), both with Calvert Vaux, the Metropolitan Park System of Boston (starting in 1878)—commonly known as the "Emerald Necklace." Dates and authorship of the projects from the section "A Selected List of Olmsted Projects," in *A Clearing in the Distance. Frederick Law Olmsted and America in the 19th Century*, ed. Witold Rybczynski (New York, London, Toronto, Sydney: Scribner, 2003), 423–426.
5 Ibid., 7; APCV—Associação Portuguesa de Corredores Verdes, "*O conceito de Corredor Verde*", *APVC—Associação Portuguesa de Corredores Verdes*. www.apcverdes.org/conceito/.
6 Jack Ahern, "Greenways in the USA: Theory, Trends and Prospects," in *Ecological Networks and Greenways. Concept, Design, Implementation*, eds. Rob. H. G. Jongman and Gloria Pungetti (Cambridge: Cambridge University Press, 2004), 49.
7 Ibid., 51.
8 Ibid., 50–51.
9 Ibid., 51.
10 Ibid.
11 The *metapolis* is a notion conceived by François Ascher to describe a new type of territory generated by metropolis, composed of heterogeneous spaces, with no precise limits and characterized by dispersion, places of mobility greatly dependent on the international urban system. François Ascher, *Metapolis: Acerca do futuro da cidade* (Oeiras: Celta Editora, 1998).
12 Rewarded with the *Sir Geoffrey Jellicoe Award* in 2013 for his remarkable landscape architecture works and major contribution for landscape and environmental issues. IFLA—The International Federation of Landscape Architects, "IFLA 2013 SGJA Awardee Gonçalo Ribeiro Telles," *IFLA—The International Federation of Landscape Architects*, http://iflaonline.org/awards/sir-geoffrey-jellicoe-award/past-sgja-awardees/2013-sgja/.
13 Spiro Kostof, *The City Assembled: The Elements of Urban form Through History* (London: Thames & Hudson, 1992), 165–172; Jellicoe and Jellicoe, *The Landscape of Man*, 260–77; Enric Batlle, *El jardín de la metropoli. Del Paisaje romántico al espacio libre para una ciudad sostenible* (Barcelona/Mexico/Amadora: Editorial Gustavo Gili, 2011), 29.
14 Rybczynski, *A Clearing in the Distance.*
15 Jellicoe and Jellicoe, *The Landscape of Man*, 279.
16 Ahern, "Greenways in the USA: Theory, Trends and Prospects," 49.
17 *The Athens Charter*, written by Le Corbusier and anonymously first published in 1941, was a result of the 4th *Congrès International d'Architecture Moderne* (CIAM), which took place in 1933 in Athens, where the principles of modern urbanism were established. Le Corbusier, *La Charte d'Athènes* (Paris: Editions de minuit, 1957).
18 APUR—Atelier parisien d'urbanisme, *La ceinture verte de Paris au XXIe siècle. Hier, aujourd'hui, demain?* (Paris: APUR, 2013), 5–11, www.apur.org/en/study/paris-green-belt-21st-century-yesterday-today-tomorrow.
19 Manuel Castells and Peter Hall, *Technopoles of the World: The Making of 21st Century Industrial Complexes* (New York: Routledge, 1994); Manuel Castells, *The Informational*

Age: Economy, Society and Culture. Volume I—The Rise of the Network Society (Oxford: Blackwell Publishers, 1998).

20 Jellicoe and Jellicoe, *The Landscape of Man*, 320–23; Ian McHarg, "An Ecological Method for Landscape Architecture. From Landscape Architecture magazine, 1967," in *The Urban Design Reader*, eds. Michael Larice and Elizabeth Macdonald (London, New York: Routledge, 2013), https://bookshelf.vitalsource.com/books/9781136205651/epubcfi/6/118%5B%3Bvnd.vst.idref%3Dch0043%5D!/4/4/2/2%400:5.

21 Jellicoe and Jellicoe, *The Landscape of Man*, 320.

22 Donella H. Meadows et al., *Club of Rome*.

23 Thomas Sieverts, "Where We Live Now," in *Where We Live Now. An Annotated Reader*, ed. Matthew Stadler, www.suddenly.org, 2009, www.publicationstudio.biz/images/content/where_we_ live_now_ FINAL.pdf.

24 Ascher, *Metapolis*.

25 Thomas Sieverts denies the dense traditional city and proposes a new model characterized by "an urbanized country-side in an urban region, a landscape between nature and culture": the Zwishenstadt. Sieverts, "Where We Live Now," 54.

26 Ascher, *Metapolis*.

27 New territories are seen by André Corboz as *palimpsests*: complex systems of superposed networks, imposed by multiculturalism, mobility and the displacement of the centers of decision. André Corboz and Sébastien Marot, *Le Territoire comme palimpseste—et autres essais* (Paris / Besançon: Les éditions de l'imprimeur, 2001).

28 Ignasi Solà-Morales emphasizes the importance of urban voids—*terrains vagues*—left empty as a consequence of deindustrialization. Ignasi Solà-Morales, *Territorios* (Barcelona: Gustavo Gili, 2002).

29 Batlle, *El jardín de la metropoli*, 41.

30 APUR—Atelier parisien d'urbanisme, *La ceinture verte de Paris au XXIe siècle*.

31 The *Parc de La Villette* was designed by Bernard Tschumi and the *Promenade Plantée* by Philippe Mathieux and Jacques Vergely.

32 Cité de l'architecture & du patrimoine, "Jean-Claude-Nicolas Forestier. Biographie," *Portraits d'architectes*, 2016. www.citechaillot.fr/ressources/expositions_Virtuelles/portraits_architectes/biographie_FORESTIER-J-C-N.html.

33 José Augusto França, *Lisboa, Urbanismo e Arquitectura* (Lisbon: Colecção Biblioteca Breve, Instituto de Cultura e Língua Portuguesa—Ministério da Educação, 1989), 51.

34 Françoise Le Cunff, "Do Passeio Público ao Parque da liberdade," *Revista Camões*, no. 15/16 (2003), 179. http://cvc.instituto-camoes.pt/conhecer/biblioteca-digital-camoes/revistas-e-periodicos/revista-camoes/revista-no15-16-marques-de-pombal/1492–1492/file.html.

35 França, *Lisboa, Urbanismo e Arquitectura*, 51; Le Cunff, "Do Passeio Público ao Parque da liberdade," 56.

36 França, *Lisboa, Urbanismo e Arquitectura*, 65.

37 Ibid., 64–72. The works for *Avenida da Liberdade* started in 1879.

38 Ana Tostões, "Keil. Arquitecto dos jardins e parques de Lisboa. A história de um trabalhador humanista," in *Keil do Amaral. O Arquitecto e o Humanista*, CML—Câmara Municipal de Lisboa (Lisbon: CML, 1999), 84–88; Le Cunff, "Do Passeio Público ao Parque da liberdade,"183–185. The new park was only accomplished in the 30s of the next century, following the design of Keil do Amaral.

39 Ibid., 80; CML—Câmara Municipal de Lisboa, *Guia do Parque Florestal de Monsanto* (Lisbon: CML, 2011), 17; Teresa Camara and Mafalda Jácome, "Parque Florestal de Monsanto," *DGPC—Direção-Geral do Património Cultural. SIPA—Sistema de Informação para o Património Arquitetónico* (2013), www.monumentos.pt/Site/APP_PagesUser/SIPA.aspx?id=34085

40 João Sousa Morais and Filipa Roseta, *Os Planos da Avenida da Liberdade e seu prolongamento* (Lisboa: Livros Horizonte, 2005), 27–29.

41 Jean-Louis Cohen, "Les Olmsted et la métropole moderne," in *Le paysage en préalable. Michel Desvigne. Grand Prix de l'urbanisme 2011. Joan Busquets, Prix spécial*, ed. Ariella Masboungi (Marseille: Editions Parenthèses, 2011); Cité de l'architecture & du patrimoine, "Jean-Claude-Nicolas Forestier. Biographie."

42 Morais and Roseta, *Os Planos da Avenida da Liberdade e seu prolongamento*, 30–33.
43 Tostões, "Keil. Arquitecto dos jardins e parques de Lisboa. A história de um trabalhador humanista," 80; CML—Câmara Municipal de Lisboa, *Guia do Parque Florestal de Monsanto*, 17.
44 Teresa Barata Salgueiro, *Lisboa, Periferia e Centralidades* (Oeiras: Celta Editora, 2001), 47.
45 Tostões, "Keil. Arquitecto dos jardins e parques de Lisboa. A história de um trabalhador humanista," 79–84; CML—Câmara Municipal de Lisboa, *Guia do Parque Florestal de Monsanto*, 17–18.
46 Jellicoe and Jellicoe, *The Landscape of Man*, 302.
47 Batlle, *El jardín de la metropoli*, 35.
48 Tostões, "Keil. Arquitecto dos jardins e parques de Lisboa. A história de um trabalhador humanista," 80–81.
49 Salgueiro, *Lisboa, Periferia e Centralidades*, 49.
50 Ibid., 51.
51 Ibid., 49.
52 Ibid., 53.
53 Portugal entered the European Economic Community—EEC in 1985 (later, European Union—EU).
54 CML—Câmara Municipal de Lisboa, *Guia do Parque Florestal de Monsanto*, 19.
55 Salgueiro, *Lisboa, Periferia e Centralidades*, 59.
56 CML—Câmara Municipal de Lisboa, *Guia do Parque Florestal de Monsanto*, 19.
57 Sieverts, "Where We Live Now," 22.
58 Ascher, *Metapolis*.
59 Batlle, *El jardín de la metropoli*.
60 EC—European Commission, "Green Paper on Urban Environment: Communication from the Commission to the Council and Parliament," *EUR-lex*. 1990, http://eur-lex.europa.eu/legal-content/EN/TXT/PDF/?uri=CELEX:51990DC0218&qid=1432048192317&from=EN.
61 Salgueiro, *Lisboa, Periferia e Centralidades*, 83.
62 Gonçalo Ribeiro Telles, coord., *O Plano Verde de Lisboa. Componente do Plano Director Municipal de Lisboa* (Lisbon: Edições Colibri, 1997).
63 Ahern, "Greenways in the USA: theory, trends and prospects," 38.
64 Telles, *O Plano Verde de Lisboa*, 15. Translation by the author.
65 Ibid., 19–21.
66 Ibid., 92. This proposal comes in line with the Plan for the Revitalization of Monsanto (PORM) first published in 1990 and revised in 2001.
67 In Portuguese: *Corredor Verde*.
68 Telles, *O Plano Verde de Lisboa*, 95–97; Gonçalo Ribeiro Telles, José Adrião and Pedro Pacheco, "Entrevista. Gonçalo Ribeiro Telles. José Adrião e Pedro Pacheco. Lisboa, Dezembro 1999," *Documentos de Arquitectura. DA03* (Lisbon: A.E.D.A—Associação de Estudos Documentos de Arquitectura, 2000), 42.
69 Manuela Raposo Magalhães, *A Arquitectura Paisagista. Morfologia e Complexidade* (Lisbon: Editorial Estampa, 2001, 457.
70 João Rocha e Castro, "O Corredor Verde para Monsanto. Do Conceito à Realidade . . . Um processo de 35 anos," *Rossio. Estudos de Lisboa*, *CML—Câmara Municipal de Lisboa*, no. 02 (2013), 212–226, https://issuu.com/camara_municipal_lisboa/docs/rossio.estudos_de_lisboa_02_issuu.
71 Batlle, *El jardín de la metropoli*, 39.
72 Telles, Adrião and Pacheco, "Entrevista. Gonçalo Ribeiro Telles. José Adrião e Pedro Pacheco. Lisboa, Dezembro 1999," 41.
73 Gonçalo Ribeiro Telles, "O Corredor Verde," *Rossio. Estudos de Lisboa*, *CML—Câmara Municipal de Lisboa*, no. 02 (2013), 206. https://issuu.com/camara_municipal_lisboa/docs/rossio.estudos_de_lisboa_02_issuu.
74 Ibid., 207.
75 Ascher, *Metapolis*; Sieverts, "Where We Live Now."
76 Ahern, "Greenways in the USA: theory, trends and prospects," 38.
77 Telles, "O Corredor Verde."

78 João Rocha e Castro, "O Corredor Verde para Monsanto. Do Conceito à Realidade . . . Um processo de 35 anos."
79 CML—Câmara Municipal de Lisboa, *Guia do Parque Florestal de Monsanto.*
80 Ahern, "Greenways in the USA: theory, trends and prospects," 45, on the relevance of greenways in contemporary urban areas.
81 Ariella Masboungi, ed., *Le Paysage en préalable. Le paysage en préalable. Michel Desvigne. Grand Prix de l'urbanisme 2011. Joan Busquets, Prix spécial* (Marseille: Editions Parenthèses, 2011). Desvigne was awarded the French *Grand prix de l'urbanisme* in 2011.
82 Michel Desvigne, "Le Paysage en préalable," in *Le paysage en préalable. Michel Desvigne. Grand Prix de l'urbanisme 2011. Joan Busquets, Prix spécial*, ed. Ariella Masboungi (Marseille: Editions Parenthèses, 2011), 54–59.
83 Bernard Debarbieux, "Du paysage magnifié à l'empaysagement," *National Research Programme NRP 48—Landscape and Habitats of the Alps*, 2005, www.nrp48.ch/index.html
84 CE—Council of Europe, "European Landscape Convention," *CE—Council of Europe Portal* (2000), http://conventions.coe.int/Treaty/EN/Reports/Html/176.htm.
85 Ahern, "Greenways in the USA: theory, trends and prospects," 45.
86 Batlle, *El jardín de la metropoli*, 48–49.

References

Ahern, Jack. "Greenways in the USA: Theory, Trends and Prospects." in Rob. H. G. Jongman and Gloria Pungetti (eds.), *Ecological Networks and Greenways: Concept, Design, Implementation*, 34–55. Cambridge: Cambridge University Press, 2004.

APCV—Associação Portuguesa de Corredores Verdes. "*O conceito de Corredor Verde.*" *APVC—Associação Portuguesa de Corredores Verdes.* retrieved on June 5, 2016 from www.apcverdes.org/conceito/.

Ascher, François. *Metapolis: Acerca do futuro da cidade.* Oeiras: Celta Editora, 1998.

APUR—Atelier parisien d'urbanisme. *La ceinture verte de Paris au XXIe siècle. Hier, aujourd'hui, demain?* Paris: APUR, 2013. retrieved on July 16, 2016 from www.apur.org/en/study/paris-green-belt-21st-century-yesterday-today-tomorrow.

Batlle, Enric. *El jardín de la metropoli. Del Paisaje romántico al espacio libre para una ciudad sostenible.* Barcelona/Mexico/Amadora: Editorial Gustavo Gili, 2011.

Camara, Teresa, and Mafalda Jácome. "Parque Florestal de Monsanto." *DGPC—Direção-Geral do Património Cultural. SIPA—Sistema de Informação para o Património Arquitetónico.* 2013. retrieved on May 18, 2014 from www.monumentos.pt/Site/APP_PagesUser/SIPA.aspx?id=34085.

Castells, Manuel, and Peter Hall. *Technopoles of the World: The Making of 21st Century Industrial Complexes.* New York: Routledge, 1994.

Castells, Manuel. *The Informational Age: Economy, Society and Culture. Volume I—The Rise of the Network Society.* Oxford: Blackwell Publishers, 1998.

Castro, João Rocha e. "O Corredor Verde para Monsanto. Do Conceito à Realidade . . . Um processo de 35 anos." *Rossio. Estudos de Lisboa, CML—Câmara Municipal de Lisboa* no. 02 (2013): 212–226. retrieved on June 5, 2016 from https://issuu.com/camara_municipal_lisboa/docs/rossio.estudos_de_lisboa_02_issuu.

CE—Council of Europe. "European Landscape Convention." *CE—Council of Europe Portal.* 2000. retrieved on March 02, 2011 from http://conventions.coe.int/Treaty/EN/Reports/Html/176.htm.

Cité de l'architecture & du patrimoine. "Jean-Claude-Nicolas Forestier. Biographie." *Portraits d'architectes.* retrieved on July 12, 2016 from www.citechaillot.fr/ressources/expositions_Virtuelles/portraits_architectes/biographie_FORESTIER-J-C-N.html.

CML—Câmara Municipal de Lisboa. *Guia do Parque Florestal de Monsanto.* Lisbon: CML, 2011.

Cohen, Jean-Louis. “Les Olmsted et la métropole moderne.” in Ariella Masboungi (ed.), *Le paysage en préalable. Michel Desvigne. Grand Prix de l'urbanisme 2011. Joan Busquets, Prix spécial*, 35. Marseille: Editions Parenthèses, 2011.

Corboz, André, and Sébastien Marot. *Le Territoire comme palimpseste—et autres essais*. Paris /Besançon: Les éditions de l'imprimeur, 2001.

Debarbieux, Bernard. “Du paysage magnifié à l'empaysagement.” *National Research Programme NRP 48—Landscape and Habitats of the Alps*, 2005. retrieved on May 24, 2007 from www.nrp48.ch/index.html.

Desvigne. “Le Paysage en préalable.” in Ariella Masboungi (ed.), *Le paysage en préalable. Michel Desvigne. Grand Prix de l'urbanisme 2011. Joan Busquets, Prix spécial*, 32–85. Marseille: Editions Parenthèses, 2011.

EC—European Commission. “Green Paper on Urban Environment: Communication from the Commission to the Council and Parliament.” *EUR-lex*, 1990. retrieved on June 5, 2016 from http://eur-lex.europa.eu/legal-content/EN/TXT/PDF/?uri=CELEX:51990DC0218&qid=1432048192317&from=EN.

França, José Augusto. *Lisboa, Urbanismo e Arquitectura*. Lisbon: Colecção Biblioteca Breve, Instituto de Cultura e Língua Portuguesa—Ministério da Educação, 1989.

IFLA—The International Federation of Landscape Architects. “IFLA 2013 SGJA Awardee Gonçalo Ribeiro Telles.” *IFLA—The International Federation of Landscape Architects*. retrieved on July 13, 2016 from http://iflaonline.org/awards/sir-geoffrey-jellicoe-award/past-sgja-awardees/2013-sgja/.

Jellicoe, Geoffrey, and Susan Jellicoe. *The Landscape of Man: Shaping the Environment From Prehistory to the Present Day*. London: Thames and Hudson, 1995.

Kostof, Spiro. *The City Assembled: The Elements of Urban Form Through History*. London: Thames & Hudson, 1992.

Corbusier, Le. *La Charte d'Athènes*. Paris: Editions de minuit, 1957.

Le Cunff, Françoise. “Do Passeio Público ao Parque da liberdade.” *Revista Camões* no. 15/16 (2003): 179–186. retrieved on July 6, 2016 from http://cvc.instituto-camoes.pt/conhecer/biblioteca-digital-camoes/revistas-e-periodicos/revista-camoes/revista-no15-16-marques-de-pombal/1492–1492/file.html.

Little, Charles E. *Greenways for America*. Baltimore/London: The Johns Hopkins University Press, 1995.

Magalhães, Manuela Raposo. *A Arquitectura Paisagista. Morfologia e Complexidade*. Lisbon: Editorial Estampa, 2001.

Masboungi, Ariella (ed.). *Le Paysage en préalable. Le paysage en préalable. Michel Desvigne. Grand Prix de l'urbanisme 2011. Joan Busquets, Prix spécial*. Marseille: Editions Parenthèses, 2011.

McHarg, Ian. “An Ecological Method for Landscape Architecture. From Landscape Architecture Magazine, 1967.” in Michael Larice and Elizabeth Macdonald (eds.), *The Urban Design Reader*. London/New York: Routledge, 2013. retrieved on December 16, 2015 from https://bookshelf.vitalsource.com/books/9781136205651/epubcfi/6/118%5B%3Bvnd.vst.idref%3Dch0043%5D!/4/4/2/2%400:5.

Meadows, Donella H., Dennis L. Meadows, Jørgen Randers, and William. W Behrens. *Club of Rome. The Limits to Growth. A Report for the Club of Rome's Project on the Predicament of Mankind*. New York: Universe Books, 1972.

Morais, João Sousa, and Filipa Roseta. *Os Planos da Avenida da Liberdade e seu prolongamento*. Lisboa: Livros Horizonte, 2005.

Oxford Reference. “Parkway.” *Oxford Reference*. Oxford University Press. retrieved on May 29, 2016 from www.oxfordreference.com/view/10.1093/oi/authority.20110803100307275?rskey=3hZjWP&result=4.

Rybczynski, Witold. *A Clearing in the Distance. Frederick Law Olmsted and America in the 19th Century*. New York/London/Toronto/Sydney: Scribner, 2003.

Salgueiro, Teresa Barata. *Lisboa, Periferia e Centralidades*. Oeiras: Celta Editora, 2001.

Sieverts, Thomas. "Where We Live Now." in Matthew Stadler (ed.), *Where We Live Now: An Annotated Reader*, 21–81. retrieved from www.suddenly.org, 2009. retrieved on June 5, 2016 from www.publicationstudio.biz/images/content/where_we_live_now_FINAL.pdf.

Solà-Morales, Ignasi. *Territorios*. Barcelona: Gustavo Gili, 2002.

Telles, Gonçalo Ribeiro (coord.). *O Plano Verde de Lisboa. Componente do Plano Director Municipal de Lisboa*. Lisbon: Edições Colibri, 1997.

Telles, Gonçalo Ribeiro, José Adrião, and Pedro Pacheco. "Entrevista. Gonçalo Ribeiro Telles. José Adrião e Pedro Pacheco. Lisboa, Dezembro 1999." in *Documentos de Arquitectura. DA03*, 30–63. Lisbon: A.E.D.A—Associação de Estudos Documentos de Arquitectura, 2000.

Telles, Gonçalo Ribeiro. "O Corredor Verde." *Rossio. Estudos de Lisboa, CML—Câmara Municipal de Lisboa* no. 02 (2013): 206–211. retrieved on June 5, 2016 from https://issuu.com/camara_municipal_lisboa/docs/rossio.estudos_de_lisboa_02_issuu.

Tostões, Ana. "Keil. Arquitecto dos jardins e parque de Lisboa. A história de um trabalhador humanista." in CML—Câmara Municipal de Lisboa, *Keil do Amaral. O Arquitecto e o Humanista*, 79–90. Lisbon: CML, 1999.

13 Weather, Science and Cinema

In Conversation With Conor McFeely

Conor McFeely and Gabriel N. Gee

Artist Conor McFeely was born in Derry, Northern Ireland, where he lives and works today. His work incorporates a wide range of forms and media, from the ready-made to sculpture and installation, as well as photography, video and audio. A chief characteristic of his work is the fracturing and manipulation of 'material' in the search for new relationships. Often read as multi-layered, many of his works have been driven by ruminations on the nature of free will, choice and autonomy. Contexts and source materials demonstrate his interests in the history of counter culture, literature and social contexts. McFeely's work is historically mindful, with references ranging from 20th-century global political history, Northern Ireland's troubled legacies and landscapes in the second half of the 20th century, as well as scientific and epistemological histories, which are of particular relevance to this volume.[1]

Our discussion begins with *The Weathermen Projects* (2013–2014), a series of works whose departure point are the counter-revolutionary US group *the weathermen*, who were active in the 1960s. McFeely's works represent the collisions between local urban radical political action and macroscopic cosmological forces at the turn of the 1960s. The discussion turns then to the role of science and scientists in the shaping of human and earthly environments in the post-nuclear age. In *The Case of the Midwife Toad* (2007), McFeely revisited the theories, anathema and implications of Paul Kammerer's experimentations and theories regarding the inheritance of acquired characteristics, which opens a broader reflection on human agency in the face of natural determinism. We continue with a discussion about McFeely's *The Prisoner's Cinema* (2015), a series of works which experiment with the hallucinatory visions experienced by people kept in long-term darkness. The closing discussion further explores questions about the essence of human nature in the age of cinema, television and portable digital devices.

1 The Weathermen Projects

G.G: In the *Weathermen Projects*, cosmological forces mix with microscopic influences. History unfolds on different scalar levels, and violence lurks in man as much as in nature. In his *Subterranean Homesick Blues*, 1965, Bob Dylan sings "You don't need a weatherman, to know which way the wind blows." Indeed, as Dylan suggests, the weather is irremediably political, carrying the scent and threat of imminent change. The *Weathermen Projects* evoke the spirit of 1960s liberation movements, which could turn violent, as in the case of the eponymous US revolutionary group. Their subterranean plotting envisioned a form of radical opposition; yet effectively it operated from a micro-scalar level, in secrecy. The opening sequence of the video film *How Irish is it?* in *The Weathermen Projects* is accompanied by the seemingly prophetic opening words of Robert Musil's great modernist novel *The Man Without Qualities*, written in the 1930s:

> There was a depression over the Atlantic, it was traveling eastward towards an area of high pressure over Russia and showed no tendency to move northwards around it. The isotherms and isotheres were fulfilling their functions. The atmosphere at temperature was in proper relation to the average annual temperature, the temperature of the coldest as well as the hottest months. The rising and setting of the sun, the moon, the phases of the moon, of Venus, of the rings of Saturn, and many other significant phenomena were all in accordance with the forecasts in the astronomical yearbooks. The water vapor in the air was at its maximal state of tension, while the humidity was minimal. In a word that characterises the facts fairly accurately, even if it is a bit old fashioned: it was a fine day in August 1913.[2]

In the video sequence, celestial views and radar waves animate the darkness of space above the earth, and specifically Derry, Northern Ireland, to which the next sequence zooms in a panoramic view of rooftops underneath the fast changing sky. The film conveys a sense of world history. Yet celestial matters also mix with housing matters. Underneath the eternal cycle of days and nights, and the rotation of the earth, there is an indication of possible explosions, which might turn the world upside down, or remain localized accidents unnoticed from superior gravitational forces. The imbrication of the cosmos with human affairs has a long history. The technological mingling of human activities with cosmological configuration has, however, taken a new dimension in the 20th century: might *The Weathermen Projects* point to an emancipation tainted with possibly apocalyptic consequences? To what extent does the progressive agenda of the sixties might inflect the demiurgic God made man?

CMF: During the process of *The Weathermen Projects* I was conscious of the rumblings of history around the planet and looking at any possible discernible patterns emerging from them. The image of the weather was employed as a common denominator linking the various references that later grew through the construction of the works. Musil's text was the starting point. It became the top layer enveloping and wrapping everything below, it is a force we are all subject to and controlled by to varying degrees. It can be read literally in the meteorological sense but is much more nuanced and alludes to the ferment and change taking place at ground level across the globe. The Weathermen movement from the sixties, the underground revolutionary group intent on overthrowing the American government, is deliberately referenced but acts as stand-in for a plethora of similarly intentioned and sometimes associated groups around the world at the same time; forecasters brewing the circumstances for possible futures in opposition to the prevailing conditions. This layering up of ideas became a device for looking at these networks. There seemed to be an implied vertical trajectory providing a shifting viewpoint that when zoomed to its outermost limit revealed some astral vision of events: clarity above the clouds perhaps. This path can be imagined rising through one of the key icons in the project, that being the pint of Guinness as a biosphere (Figure 13.1). It became a container providing a focus for meditation and speculation. It echoes classic images of the earth as pictured from outer space from the same period. The Weather Underground can be considered as an example of one of many groups in the sixties and seventies who shared a desire to

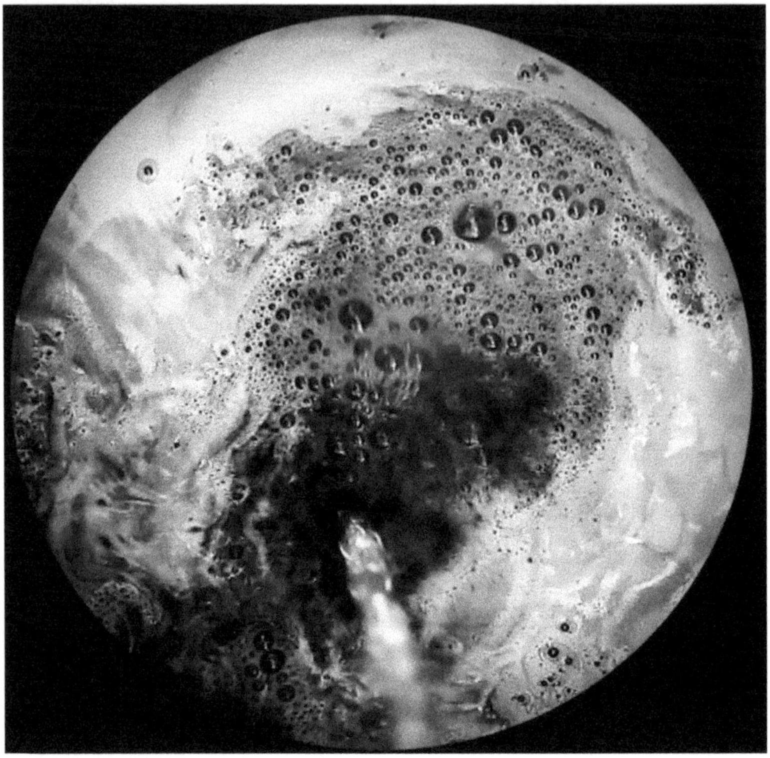

Figure 13.1 Conor McFeely, *Weathermen Projects*, 2012
courtesy of the artist

destabilize and overthrow the establishment. Akin to particles in suspensions these various factions when deposited could be seen to form their own sediment: a collective bedrock of resistance. At the nucleus of the project lay the presence of conspiratorial gatherings getting bevvied up at an undisclosed subterranean weather bureau, mapping out some desired progression towards a new world order. In the Irish context this conjured up clandestine Republican groups in snugs but it could just as easily be the RAF or Baader Meinhof in Germany, the Angry Brigade in England or the PLO in the Middle East. Multiple narratives folded in on one another. I am conscious also that growing awareness of environmental preoccupations ran concurrently with these assorted assemblies, but with a different methodology for expressing their concerns. Prominent in this category might be GAIA theorist James Lovelock whose belief in the planet as a self-regulating organism might be part of some absurdist comedy, encircling the actions of all these groups. Nature as a demiurgic God like a Goyaesque Collossus towering over all below.

2 Science and Scientists

G.G: In that respect, the realm of science and the figure of the scientist recur regularly in your work. The demiurgic God playing with Gaia's fragile atmosphere might well be a scientist, bent over an experiment in atmospheric pressure behind an isolated glass container. This figure sometimes appears as an ancestral pre-modern forefather versed in mechanical sciences and secret knowledge, the alchemist, of which the contemporary artist is possibly an heir. *The Case of the Midwife Toad* for instance, muses on the story and unsettled legacy of the biologist Paul Kammerer (1880–1926). After his suicide in 1926, and having been accused of scientific fraud, his theories concerning the 'inheritance of acquired traits' were largely dismissed. Fifty years later, the journalist Arthur Koestler wrote an eponymous publication (1973) investigating the machinery of scientific production of facts, and the machiavellian plotting that could have led Kammerer to be unjustly accused. Koestler's investigation reveals that the Austrian biologist can be seen on the contrary as a pioneer in experimentation, as well as a romantic figure, whose genial inspiration is met with corporate scorn. Metaphorically, he represents the frailty of scientific facts, which experiment and observation are meant to affirm unequivocally. His legacy encapsulates the 'politics of nature,' in which science has been increasingly asked to take part in. Beside the centrality of doubt to all scientific experimentation and outcomes, on the other hand, there is in the *Weathermen Projects* for instance, a different scientific figure who appears resolutely more chilling: a man in a white blouse standing with a stethoscope in his hand behind an abstract grid. That the world must be cured of its illnesses is a thought to which in diverse circumstance science has given a helping hand in the course of history. The evolution of science in the 20th century has transformed life and death. In your film *Inside His Master's Voice*, the master's voice evokes a failed scientific project that aimed to decode a neutrino signal (an ultralight particle) from space. The voiceover is combined with the voice of J. Robert Oppenheimer (1904–1967), the figure at the center of the development of nuclear weapons in the United States of America in the 1940s. The presence of Elizabethan collars in the film, collars that simultaneously prevent a dog from scratching a wound, and also limit vision, expose a reflection on causality: the frame within which creative actions and ideas are being channeled in history. When combined with political might, the effects can take biblical apocalyptic forms. In the words of Oppenheimer: "We knew the world would not be the same, a few people laughed, a few people cried, most people were silent." There the mood and the role of science in the second half of the 20th century seem to be set. Between the alchemist and the 20th-century professional scientist, both meddling with the laws of nature, there is a world apart, or are they in fact the two faces of the same divinity?

CMF: The scientific inquiries referred to in my work run parallel to political thought though they are not necessarily constructed by it. Kammerer's tests seemed to form the basis for a belief in our ability to control our own destiny by confirming the inheritance of acquired characteristics. Through the 'evidence' he presented it seemed it was apparently possible to engineer our future. The principle of biological self-determination highlighted in his work is a shift away from a subservient slave-like narrative account of the species. It raises complex questions about the existence of freewill, choice and their limitations. Power structures are challenged and not least that of the notion of a divine power with creative intention. Through his experiments with amphibians the idea of the grand designer appears to be a narrative we constructed to explain our evolutionary purpose rather than the other way round. Koestler might have said we were sleepwalkers until Kammerer and others shone their searchlights and woke us up to that fact that we can become authors of the story and not characters within it. His ideas were in tune with the ideology of post-revolutionary Russia where he became something of a hero as well as being at odds with the growing fascist movement in Germany. The protagonists evoked in the *Weatherman Projects* and *Inside His Master's Voice* functioned in similar ways. An artwork of course can't establish any scientific truth but can point at or direct attention to other fields of activity. *The Midwife Toad* installation became a performance of sorts driven by an interest in outsiders prepared to go to extremes despite the consequences. The "doctor" in *Weatherman Projects* represents such a shadow figure and was an amalgam of characters whose interrogations were open ended, possibly gratuitous and without moral imperative and whose results contributed little of benefit to society except to reinforce preconceptions. Kammerer's work in the field of epigenetics sits adjacent to research in eugenics favored particularly in Germany where the God-like intentions of the regime attempted to eradicate and cleanse the species of perceived flaws and weaknesses. The well documented enterprise conducted by Mengele under the auspices of the Third Reich Project was fused with Doc Benway from William Burrough's *The Naked Lunch* (1959): the latter being more interested in the particulars of his extreme procedures than any constructive results that might have arisen from them. The model of the scientist as creator is very much alive and well and leaves us asking questioning our position on methodology. The pursuit of new treatments through vivisection or the use of gene treatment to eliminate undesirable conditions and purify the species still divide opinion.

3 The Prisoner's Cinema

G.G: The sculptures and installations entitled *Phosphene Diaries* explore the phenomena of internal vision: the seeing of light without light rays actually entering the eye's retina. In the more recent project *The Prisoner's Cinema*, you again pursue this theme which furthers a reflection upon empowerment and alienation. The prisoner's cinema refers to the experience of people kept in long-term isolation in the dark. After a while, individuals experience internal visions, a phenomenon known as the Ganzfeld effect. This work evokes a form of *interstitial* freedom to be found at the heart of incarceration. We see a man half naked in a foreshortening angle wearing dark goggles: a sci-fi adventurer who treads upon new worlds while he lies still in a horizontal position (Figure 13.2). This scene evokes the main protagonist of Chris Marker's *La jetée* (1962), a survivor of the atomic war which has destroyed Paris who is sent by his jailors on an internal quest for an escape route into the future. In the *Prisoner's Cinema* we also see footage of early cinema scenes, alluding to the early visionary of a medium which has taken, with television and screens, humanity into its claws; can the early days of experimental cinema offer an analogy to the prisoner's interstitial emancipatory vision? A time when filmic projection was unadulterated by the serial commercialization that accompanied the expansion of the industry in the booming years of the 1950s and 1960s . . . And we also see a bat, revolving cinematographically in hypnotic manner, an icon for the project in many ways, due to the animal's capacity to 'see' in the dark, through its biological sonar and echolocation; in the age of virtual fluxes, the bat might stand as a heraldic symbol of resistance suggesting the emancipation of the subject from within. *The Prisoner's Cinema* interlinks 20th- and early 21st-century human and post-human nature: screens since the 1950s have gained increasing presence in our societies, infiltrating houses, communities and urban habitats, from the streets to the couch, thereby extending humans' social networks, but also filtering and channeling our perception of the world. A return to the birth of cinema, combined

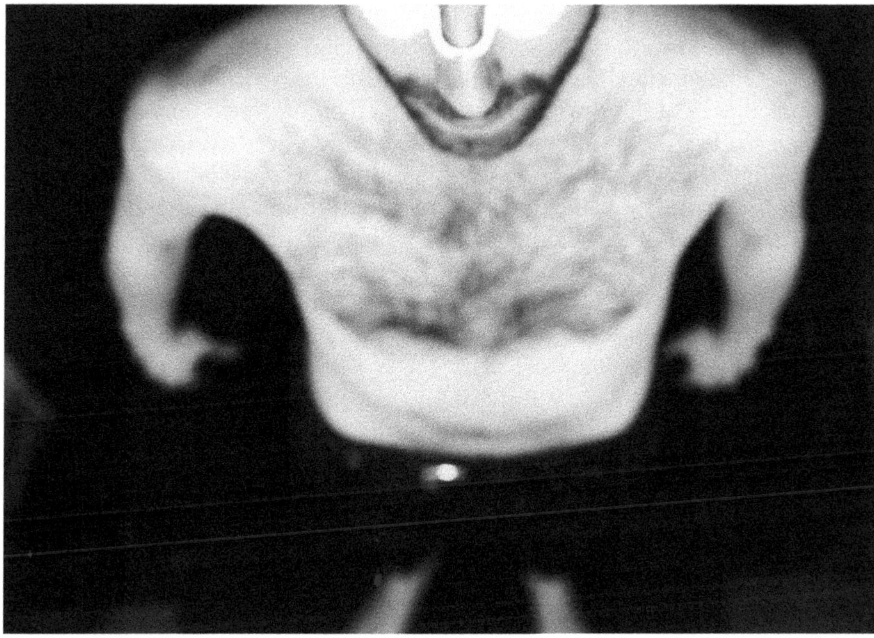

Figure 13.2 Conor McFeely, *The Prisoners' Cinema*, 2015
courtesy of the artist

with the physical phenomena of a phosphene internal vision, opens the door into a subversion of tentacular global media superstructures. Deep inside the human's physical essence, there lies the possibility of being in control of one's thoughts. From within our natural being there might be an interstitial alternate route and future shying away from a global intermedia logic that has developed in stupendous fashion in the past 50 years all across the world. In the age of spatial exploration, and of increasing interconnectedness on the earthly grounds, the *Prisoner's Cinema* thereby also questions our changing cosmological views.

CMF: One of the side effects of the expansion of interfaces like the cinema and the web into the framework of our lives has been to reduce earlier collective engagements with the mythologies of the cosmos and supplant them with signs from these new beacons. The transformative cinematic experience is one of solitary disembodiment, captured alone in the dark with sensory perception reduced to sight and sound. The loss of self is the desirable sensation and maybe the logical consequence of this contact. The perspective of the lone quest into uncharted territory in pursuit of the named and the nameless is common to most art forms with cinema playing a prominent role. Go in too deep and the apparent drunkenness experienced by deep-sea divers, nitrogen narcosis or the 'narks' awaits. *The Prisoner's Cinema* was a plunge through the screen of the retina, an invasion of the self into hypnagogic hallucination and involuntary imagery where navigation and spatial coordinates are guided sonically. Extreme predicaments such as brain injury or others involving forms of imposed restraint and sensory denial or food deprivation can dissolve body consciousness and open new gateways. This disruption of neural pathways may have a corresponding expansion, a mirror world of the interior that we could call true nature. The video sequences in *The Prisoner's Cinema* took the form of cinematic vignettes in the sense of them being short chronicles. The term vignette was used to describe a visual phenomenon in early cinema; a central area of light on the screen fringed by darkness that often came about because of lens limitations. Ironically it's a look often used consciously for effect in digital filmmaking today. Appropriated footage from 1920s cinema is mixed with my own to form a hybrid. It's challenging for us today to imagine the escape offered to audiences by the portal of early cinema. The widening of the iris and pupil on returning to the street outside viewed through mesopic vision provokes parallels with 'coming down' from an acid trip. Slowly returning and rebuilding a coherent world outside us that has been changed permanently by what happened moments before. In the final scenes of Roger Corman's *The Man With the X-ray Eyes* from 1963 we witness the central player debilitated by the effects of experimenting with a drug that allows him to see beyond normal sight, through reality. In the closing scenes, unable to cope with his revelations, we experience through his eyes his final tormented odyssey across city and landscapes of exposed infrastructures, until he stumbles into a fundamentalist meeting house to be greeted with cries of "If thine eye offends thee pluck it, pluck it out."

Notes

1 www.mentalimage.org.uk
2 Robert Musil, *The man without qualities*, Sophie Wilkins & Burton Pike (trans.), London: Picador, 2011.

14 The Postcolonial Garden City? Changing Representations of Nature in Christchurch, New Zealand

Alison Vogelaar and Brack Hale

In December 2014, *The New Yorker* published an article by Elizabeth Kolbert titled, "The Big Kill: New Zealand's Crusade to Rid Itself of Mammals."[1] The feature describes the sometimes gory, often toxic and potentially futile extremes to which New Zealanders have in recent times gone to attempt to "wipe out all mammalian predators." The article offers a nuanced exploration of the problems of postcolonial conservation in the specific context of New Zealand where every mammal (save two species of bat) is a non-native and almost every citizen is a conservationist/exterminator. Indeed, according to the author, "on a per-capita basis, New Zealand may be the most nature-loving nation on the planet. With a population of just four and a half million, the country has some four thousand conservation groups." And these groups mean business; as the coordinator of one volunteer group put it to Kolbert, "We always say that, for us, conservation is all about killing things."[2] Though the essay makes light of the mammalian carnage (and there is a lot), it does not make light of the precarious postcolonial situation in which New Zealand's fragile ecosystems find themselves, and the many questions that begs: Can we return to a 'native' ecosystem? What are the grounds for such a pursuit? And, what is native?

This essay explores the changing representations of 'nature' and 'the native' in the specific context of New Zealand by examining the evolution of two notable and markedly different representations of urban nature in Christchurch: the Christchurch Botanical Gardens and Riccarton Bush. In the context of this volume, and as you will read below, both sites are interstitial spaces—literal and symbolic grounds on which the urban and wild, natural and synthetic have come into constant contact and articulation. Our analysis, situated in the literatures of postcolonial conservation/nature studies, explores colonial and postcolonial mindsets, practices, institutions and discourses as they have transformed the ways in which nature is represented in two important public landscapes in Christchurch. New Zealand offers a poignant case study of colonial transformations of the natural world, and postcolonial responses to such transformations, for several reasons. First, New Zealand's native (pre-colonial) flora and fauna are the result of a unique set of geological processes and resulting ecosystems. The land that now makes up New Zealand split from the super-continent, Gondwanaland, 80 million years ago. Since that time, mountain ranges have formed, the land has sunk into the sea, and risen again, and ice ages have carved the landscape. These forces, as well as New Zealand's isolation from other land masses, have resulted in ecological peculiarities; indeed New Zealand has many "impossibly outdated" and "outlandishly innovative" species.[3] Second, transformations to New Zealand's natural world were more accelerated than in other landscapes, "with huge changes telescoped into one thousand years of human habitation."[4] The first settlers (Polynesians from whom the Maori descended) arrived only about 800 years ago and European colonization of New Zealand occurred rather late in the Empire (mid-19th century). Given such recent and rapid colonization, many clues still remain to what New Zealand's environment was like before human inhabitation. For example, although the moa, a giant bird, has been extinct for centuries, fossil and narrative remains tell a coherent story about the animal and its place in the ecosystem. We also have a good record of what the landscape was like before its transformation under colonial rule. A final reason New Zealand makes for an interesting case study of colonial and postcolonial transformations is its unique reactions to these colonial legacies in contemporary (postcolonial) times, which include both the Waitangi Tribunal, convened in 1975 to settle historical Maori grievances about dubious British land grabs and 'sales' under the 1840 Treaty of Waitangi and concurrent conservation efforts to 'go native,' which some worry may result in a counter-productive form of eco-nationalism that attempts to bring Maori and European settlers together through the act of 'fixing' New Zealand's nature in some imagined pre-colonial past.[5]

Colonial Science, Colonial Nature

In addition to fundamentally altering the political, economic, geographic and social fabric of much of the world, the British Empire transformed 'nature,' "creating new landscapes, new ecologies and new relations between human and non-human nature."[6] Indeed, mastery over nature was one of the fundamental features of both British and European colonial expansion. In this sense, colonization was both a material and a symbolic project. In addition to 'providing' material resources (land, energy, raw materials, labor), colonies were imagined as "a golden opportunity for retrieving some of the knowledge of nature man had had in the Garden of Eden."[7] This is not to suggest that colonial efforts and effects, even within the territory of the British Empire, were homogenous. Colonial influences on the natural world were vast, complex and highly dependent upon the unique time, place and people involved. That being said, some generalizations can be made about how European (and specifically British) colonization transformed nature both literally and symbolically.

Postcolonial theorists have astutely demonstrated the mechanics, effects and contradictions of the organization of the colonial world into 'centers' and 'peripheries'[8] with European urban sites and subjects representing the geopolitical, cultural, scientific and economic centers of the European colonization and colonial territories and subjects representing the rural, dependent, natural and extractive peripheries. While the center and periphery supposedly represented spatial and geographic realities, this binary was utterly symbolic, constructed by those in power to describe, indeed reproduce, power relations in their own terms. The oversimplification of spatial and cultural relations suggested by this binary has been productively problematized in recent years, with postcolonial scholars astutely noting the more complex exchanges and networks that in fact characterized the colonial period.[9] Even so, the imaginative force of the 'centre-periphery' binary had major bearings upon the development and locations of science and ecology in this period and is as such a useful starting point for this exploration of the lingering effects of colonialism in present-day New Zealand.

Colonial expansion and exploitation were intimately embedded in the intellectual ferment of the European Enlightenment. According to William Mark Adams and

Martin Mulligan, the "destructive, utilitarian and cornucopian view of the feasibility of yoking nature to economic gain" was a direct result of Enlightenment thinking about the natural world.[10] Both the Cartesian dualism of mind–matter (and, by extension, nature–culture) and economic models of nature as resource born in that period were central in the transformation of the natural world. This is not to glorify or overlook pre-colonial ideas about, and interactions with, the natural world (which were not always positive) but to acknowledge that the European Enlightenment constituted both a major shift in thinking about the natural world and the technological capacities to manifest such thinking on a scale heretofore unimaginable.

The "profound belief in the possibility of restructuring nature and re-ordering it to serve human needs and desires"[11] was made possible in part by scientific practice in this period. Natural history, early modern botany and what would become ecology, each intimately supported and profited from European expansion.[12] The classification and cataloging of nature (e.g the Linnaean system) prominent in the 19th century was one key way in which the natural world was rationalized, commodified and dispossessed. Drawing on pre-existing views of the colonial periphery, colonial scientific discourses and practices took possession of, institutionalized and exported biological specimens[13] in what Mastnak et al. have called "a massive operation of planting and 'displanting.'"[14] With science as its tool, colonialism promoted the naming, classification and mobilization of specimens with the aims of control and utility.

As early formal scientific institutions, botanic gardens played a significant role in the colonial expansion of the West and the growth of "an integrated world system based in industrialized cores which extract raw materials from rural peripheries."[15] The meanings, uses and functions of botanic gardens transformed over the course of colonial expansion and retreat (from medicinal, economic, educational and aesthetic). As centers of plant importation and exchange as well as knowledge production in the name of imperial economic development, botanic gardens reinforced the privileging of the "urban, lettered, male authority over the whole of the planet."[16] They were also the site where the colonial pattern of sending scientific knowledge of nature from the periphery to the metropole—so that its value could be assessed, amassed and converted into science—was institutionalized and perfected by scientists.[17]

The widespread plundering of nature during European colonization (made possible in part by/through economic and scientific conceptions of nature as resource) did not, however, go unnoticed. Indeed, "during the second half of the 19th century, the destructive power of human activities was widely appreciated in North America and Europe."[18] It was during this period that the idea of conservation emerged from the field of ecology.[19] The conservation organizations emerging in this period, both in colonial 'centers' and 'peripheries,' unsurprisingly "reflected the ideologies of nature that grew out of utilitarian and reductionist 'natural sciences.'"[20] So while the conservation movement reflected a "social reaction against technology, industrialization and growth-oriented development," it was also deeply imbued with the European Enlightenment values that drove global colonial development.[21] National parks and protected areas were one significant manifestation of the emergent conservation/preservation mindset in the late 19th century. Notably, reserves, parks, refuges and sanctuaries began first in the 'peripheries' (colonies and former colonies) where environmental degradation was more visible (given the transformations that took place in very short periods) and more emotionally valenced (as a loss of something perhaps sacred or Edenic).

Contemporary conservation practices, and the understandings of nature that underpin them, remain deeply embedded in the colonial legacies and contradictions discussed above. The following sections explore these legacies and contradictions in two urban natures in New Zealand: Christchurch Botanical Gardens and Riccarton Bush.

'Founding' Christchurch

Prior to the arrival of the Waitaha, the first Māori group to occupy the area, the landscape of today's Christchurch region (Ōtautahi) was predominantly shaped by occasional fires and floods.[22] The vegetation consisted of pockets of hardwood-podocarps forests (dominated by kahikatea, tōtara and mataī), grasslands, and flax and sedge wetlands along rivers and the coast.[23] Māori settlers (first the Waitaha, followed by the Ngati Mamoe, and finally the Ngāi Tahu) changed the landscape somewhat, harvesting native bird species (some to extinction), introducing exotic species, and executing burns in some areas.[24] The fires reduced forest cover, particularly in the eastern areas of the South Island.[25] Over hundreds of years, however, Māori culture became intricately connected with the natural world coexisting more sustainably with the environment than did later European settlers.[26] The Māori used specific areas (known as *mahinga kai*) around Christchurch for food and resource gathering. There were also several Māori settlements in the area, including one along the Avon River (Ōtākaro) in what is now the Central Business District not far from the Botanic Gardens and a Ngāi Tahu village at Riccarton (Pūtaringamotu), which were also an important *mahinga kai* then.[27]

The Treaty of Waitangi, signed by the British Crown and Māori chiefs in 1840, marked the beginning of British settlement of New Zealand and instigated the establishment of what would become the South Island's largest city, Christchurch. The land in the area was originally purchased from the local Ngai Tahu by the colonial government with a provision for Ngai Tahu access to the important food collecting areas,[28] an agreement that was later rescinded and became a focal grievance point in the Waitangi Tribunal. Prior to the Canterbury Association's[29] acquisition of two million acres in the Christchurch area, the Deans, early settlers on the South Island from Scotland, held some of land in this region, including both Pūtaringamotu (Riccarton Bush) and the area that would become Hagley Park.[30] Although they lost most of the land to this British sale, they retained some around Pūtaringamotu, which eventually became Riccarton Bush, one of the focuses of this chapter.

The first group of settlers with the Canterbury Association arrived in December 1850 at which point the English alteration of the landscape began. Because the plants and animals unique to New Zealand were more foreign to settlers than those in many other colonies, settlers made substantial efforts to import and plant European species. This practice was in line with the common "displacement theory" of that time,[31] which contended that European species were more evolved and thus would/should eventually replace indigenous species. In addition to providing a sense of aesthetic familiarity, settlers also knew more about managing species from their homelands. Just as the Deans imported seeds from their native Scotland in the early 1840s (Roche 1984), Canterbury settlers brought seeds from England during the 1850s. This practice, combined with other colonial land uses, led to a drastic decline in the number of native species recorded in the area from 88 in 1864 to 35 in 1918.[32]

Making the Garden City: Christchurch Botanic Gardens and Riccarton Bush

Christchurch, founded during the second wave of the British Empire and in the thick of the Industrial Revolution, was in part a representational project that aimed "to promote a special colony in New Zealand, an English society free of industrial slums and revolutionary spirit, an ideal English society."[33] A Garden City. Though the garden city movement would not take off until the late 19th century, the development of Christchurch was undoubtedly undergirded both by dystopian fears in that period of urban decay and unrest, as well as, utopian visions of re-creating a Garden of Eden in the 'untouched' and untainted colonial peripheries. The Christchurch Botanic Gardens and Riccarton Bush were both manifestations of these colonial fears and dreams.

Both the Botanic Gardens and Riccarton Bush are situated within and around Hagley Park (Figure 14.1), a park whose purpose was both as a physical buffer between the suburbs and the Deans' farmland to the west and a recreational space for the city's European inhabitants. Hagley Park also served as a campground for the first formal English settlers and within 12 months of their arrival, historical documents report the landscape having "a very English air."[34] Named after the country estate of an important English family in early Christchurch, the park's name symbolized its eventual English appearance. Within the original park grounds was an area set aside as the Government Domain and it is within this area that the Botanic Gardens would develop. The Domain housed an early nursery for English plants[35] that provided new trees for Hagley Park and the city. The dense native riparian vegetation was removed from the banks of the Avon,[36] and tree belts and shrubs were planted along the boundaries of the park using primarily European species, such as oaks, elms, ashes and chestnuts. In 1864, Hagley Park became the site of an "acclimatisation" garden created by the Acclimatisation Society. The garden was the source of over 700,000 English trees planted throughout the region in the late 1800s.[37] Acclimatisation societies,[38] whose manifest purpose was to make the environments more like Britain by introducing British flora and fauna, were common in Australia and New Zealand in the late 1800s.[39]

The original plans for Christchurch set land aside for a botanical garden.[40] Unlike earlier botanical gardens, especially in European centers, the Christchurch Botanic Gardens served primarily exhibition purposes, showcasing exotic species (often of European origin), rather than indigenous ones.[41] As a project and product of the British Crown, the Botanic Gardens were inspired and maintained by European settlers and scientists for settler purposes. While we do know that the Māori provided knowledge to the European settlers about the native flora, we also know that it was often neglected and always "translated into forms comprehensible to Western botanical science."[42] Māori participation in, and use of, the Botanic Gardens is undocumented.

While the gardens showcased plant specimens from around the world, the grand entrance was, and remains, decidedly colonial in spirit, design and substance, as the 2013 Conservation Plan articulates clearly:

> there is a strong sense of visual containment achieved by a semi-closed and closed canopy of predominantly Northern Hemisphere broad-headed deciduous trees and impressive sized conifers . . . These give the Gardens a high degree of time depth, which is further reinforced by late nineteenth-century and early twentieth-century

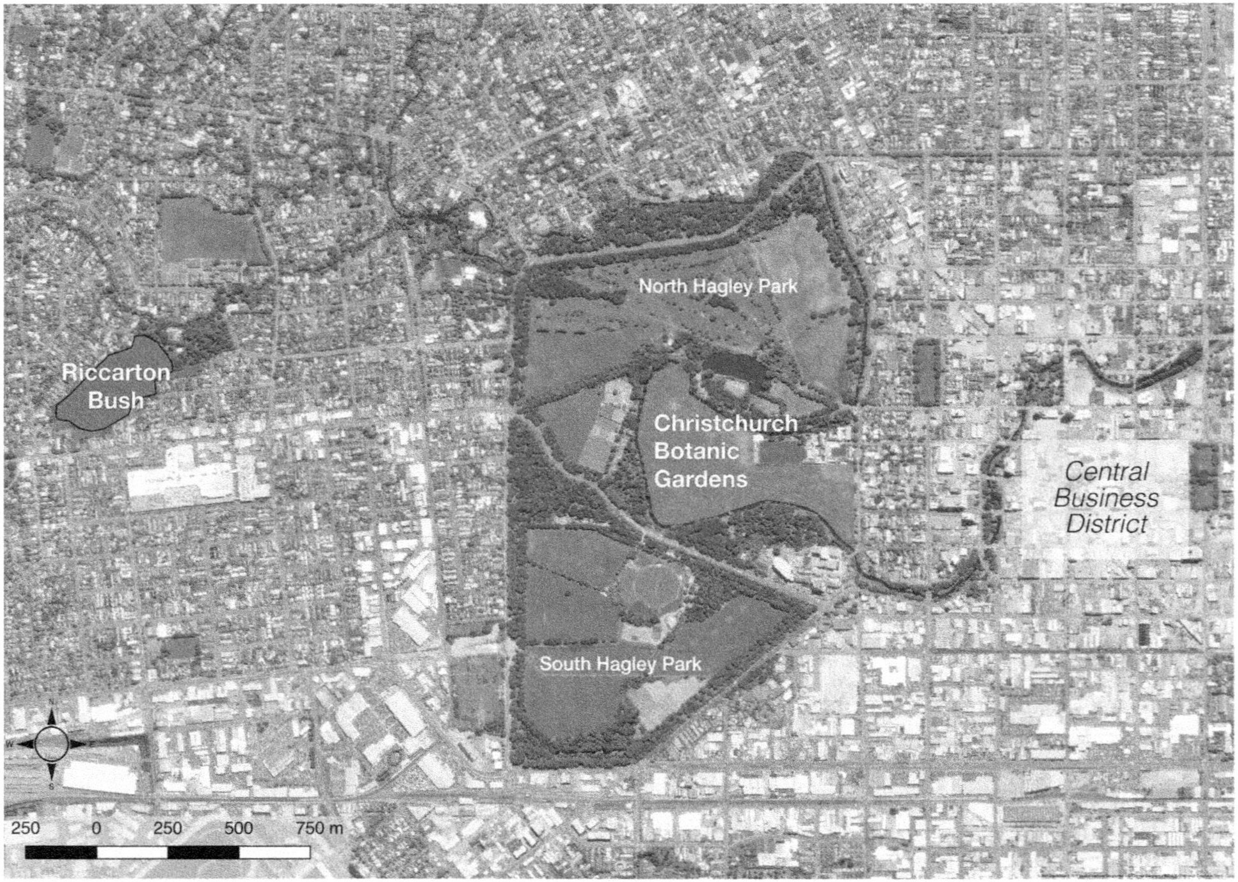

Figure 14.1 Map of Hagley Park, 2017 (B. Hale)

heritage fabric and surviving examples of the Victorian-era design principles used in the early layout of the grounds.[43]

Upon entering the present-day gardens at Hereford Street, visitors are presented with lawns lined by majestic old trees, many of which were planted as commemorations honoring the visit of an important colonial guest or event. The first commemorative tree, an English oak, was planted in 1863 to commemorate the marriage of Prince Albert Edward of the United Kingdom to Princess Alexandra of Denmark;[44] this event is now used to mark the official 'birth' of the Gardens. Indeed, most of the trees that line the entrance are European or North American species, such as oaks, beech and redwoods. English species were selected in the early development such that they might "reinforce the connection between colonial Christchurch and England and, in the case of oak plantings, display an ongoing allegiance to the Crown and the mother country."[45] Of the 32 surviving commemorative trees listed in the 2013 plan, only five are indigenous New Zealand species: two kauri, two rimu and one totara[46]—poignantly the planting of native trees only began after New Zealand achieved Dominion status within the British Empire (the two kauri[47] were first planted in the 1920s; the rimu and totara date from the 1960s and early 1970s).[48] Just as Dominion status changed New Zealand's formal political relationship with the Crown, so too we see a shift in the treatment of nature.

Despite the colonial impulses to create an English setting, there was some early interest in featuring native New Zealand species. In 1864, Joseph Armstrong, a local botanist who had immigrated to the area from Northern England, identified and recorded 88 native species present within Hagley Park.[49] In 1875, Armstrong and his father (who was Head Gardener at the time), developed an area in the Gardens called the New Zealand Arboretum which was planted with New Zealand tree species. A later curator, Young, also planted a New Zealand Garden, which lasted from 1910–1927.[50] There were also a few native plants included in some of the plantings within the Domain and Park. However, the native plantings generally served more practical (rather than aesthetic or educational) purposes as they were needed to provide food for the native birds still residing in the area and to produce seeds for export to other parts of the Empire.

Also central to the representation of a garden city at Christchurch, was the development of Riccarton Bush (Pūtaringamotu), a large natural area just a few kilometers to the west of the Botanic Gardens. The original floodplain forest patch in this area measured about 30 ha. before the arrival of Europeans.[51] The property was settled in the early 1840s by the Deans brothers of Scotland. Consistent with the times, the Deans imported and planted Scottish trees, which they believed were biologically superior to, and thus of higher economic value than, the native species. Although the landscape immediately surrounding the Deans' main house and the entrance to Riccarton Bush still is characteristically English in both design and substance, approximately 8 ha of the Deans' land represents the original native vegetation. This land was eventually protected at the dying request of John Deans in 1854 and declared a natural reserve in 1914 with the Riccarton Bush Act when the land was officially donated to Canterbury.[52] Despite its protected status, the land experienced significant impacts. Jane Deans, John's widow, reported that fallen timber was sometimes removed for fencing.[53] Additionally she planted non-native ash and oak around the borders of and in forest gaps. The area "was treated like an English woodland" in that grassy areas were mown and natural 'litter' and debris were removed from the forest floor, practices that continued until 1974.[54] Upon establishment of the reserve in 1914, the area was closed to the public for three years so that proper fences could be built and walking trails established. As the area around the Bush shifted from agricultural to suburban, so did incursions into the property in the form of new informal walking paths and disposal of garden waste in the bush. The area was also subject to the effects of invasive plants and animals as these species did not 'obey' the bush's legal boundaries and fences. Nonetheless, in contrast to Hagley Park and the Botanic Gardens, the core area of Riccarton Bush was never designed or truly managed as an English garden or setting and thus retained the base of a natural, albeit heavily impacted, landscape and ecosystem; what Given and Meurk refer to as a "primary remnant."[55] It does not mean that the area does not have the presence of non-native species, but rather that the original composition of the community has remained relatively intact.

Representing a Postcolonial Garden City

Pinning down a specific historical moment when the Christchurch region shifted from colonial to postcolonial is no straightforward task. New Zealand's 'independence' from the British Crown did not occur in one fell swoop, but rather was gained in many small steps and is complicated by the fact that the nation is constituted not just by European settlers but also by its original human settlers, the Māori. In this sense, New Zealand faced, and continues to face, external and internal colonialism. Nevertheless, the 1970s constitute an important moment in this transformation. The Māori 'Renaissance' and modern protest movement in particular are credited with profound shifts in attitudes towards, and policies around, language, culture, land rights and landscape in New Zealand. Most significantly, this moment saw the establishment of the Waitangi Tribunal in 1975, a permanent commission of inquiry that began revisiting and, in some cases, righting Māori grievances associated with the implementation of the 1840 Treaty of Waitangi.[56] The passage of the Local Government Act in 1974 also marked an important transformation in land management (among other things), as it transferred power to regionally elected councils whose boundaries followed watersheds rather than human-created boundaries.[57] In terms of the environment, the growth of the fields of conservation biology and invasion ecology in the 1970s and 1980s transformed how scientists, and eventually publics, saw and valued native biodiversity.[58] In New Zealand, as elsewhere, this manifested in the proliferation of conservation organizations, policies and practices privileging and preserving 'the native' and 'the wild.'[59]

Christchurch Botanic Gardens: Constructing Natural Heritage

While Christchurch Botanic Gardens is a fundamentally colonial project, the uses of and meanings associated with it have transformed in postcolonial times. Whereas botanic gardens were once the center of colonial science and knowledge production, they increasingly serve hybrid purposes. In recent times, botanic gardens have become a central site for the construction, representation, commodification and management of *natural heritage*, representing an undoubtedly postcolonial take on botanic gardens. Whereas the Gardens once functioned as a means to preserve the cultural heritage of the places from which colonizers had come, the postcolonial period has seen botanic gardens transform into sites for the preservation of natural heritage. The idea of protecting heritage on a global scale emerged from the horrific destruction of cultural monuments and artifacts associated with World War I, and is rooted more generally in 20th-century 'Western' responses "to the changing relationship between people and objects and between people and places as the camera (and satellites), the car (and airplane) and the computer (and the web) have radically altered how people in Westernized, industrial and post-industrial societies engage with their environment and with each other."[60] In this sense, emerging forms of heritage are *glocal*, motivated by, and involving both, global and local interests and purposes.

World heritage was made official on November 16, 1972 when the "Convention Concerning the Protection of the World Cultural and Natural Heritage" was adopted by the General Conference of UNESCO. The convention merged two separate movements at that time: one focusing on the preservation of cultural sites and the other on the conservation of natural sites.[61] The international recognition and protection of cultural heritage sites was first articulated in 1959 when the governments of Egypt and Sudan requested UNESCO's help in the protection of sites and monuments endangered by the development of the Aswan High Dam in the Nile Valley. The Valley was ultimately flooded but not before UNESCO pulled together (50+ nations contributed) USD 80 million to dismantle, remove and preserve key artifacts and monuments.[62] Unsurprisingly, it was the United States who first proposed the idea of adding natural sites to global cultural conservation efforts. As a young nation with decidedly few (European) cultural monuments, and an emerging commitment to nature conservation, the United States government called in 1965 for a "World Heritage Trust" to preserve "the world's superb natural and scenic areas and historical site for the present and the future of the entire world citizenry."[63] This, along with proposals developed in 1968 by the International Union for Conservation of Nature, was presented in 1972 to the United Nations Conference on the Human Environment in Stockholm soon after which the Convention supporting the preservation of cultural and natural heritage was adopted. The 1992 World Heritage Convention saw the official inclusion of *cultural landscapes*—"combined works of nature and of man"—in the World Heritage List.[64] Notably, two of the three World Heritage Sites in New Zealand (Te Wahipounamu and the Sub-Antarctic Islands) fall under the "natural heritage" category and the third, Tongariro National Park, was the first site to be inscribed under the 1992 revisions as a "cultural landscape" deserving of heritage status and protection.

While Christchurch Botanic Gardens is not among New Zealand's World Heritage Sites, they are a part of the general trend toward the articulation and commodification of local and national heritage on and for the world stage. Indeed, the gardens generally serve two purposes: first, they are an important site in the (controversial) construction of a postcolonial (Māori/European) national identity and second, they promote New Zealand as a tourist destination. Indeed, according to Ginn, the Botanic Gardens receive over 550,000 tourists each year.[65] Postcolonial shifts in the meanings and uses of the gardens are evident in two key phases. The post-Dominion gardens saw several changes to its makeup. As mentioned above, the tree lined grand entrance saw its first 'native' planting in 1928 (a kauri tree). In addition, beginning in the 1920s, the Gardens added the *Ō Tane Aotearoa, the New Zealand Icon Garden*, which is now divided into two sections: the New Zealand Bush and the Cockayne Memorial Garden (Figure 14.2). Originally planted around 1920, the Bush section contains a mix of native trees and shrubs that are supposed to mimic a natural forest.[66] The Cockayne Gardens, which pay homage to an English immigrant and early

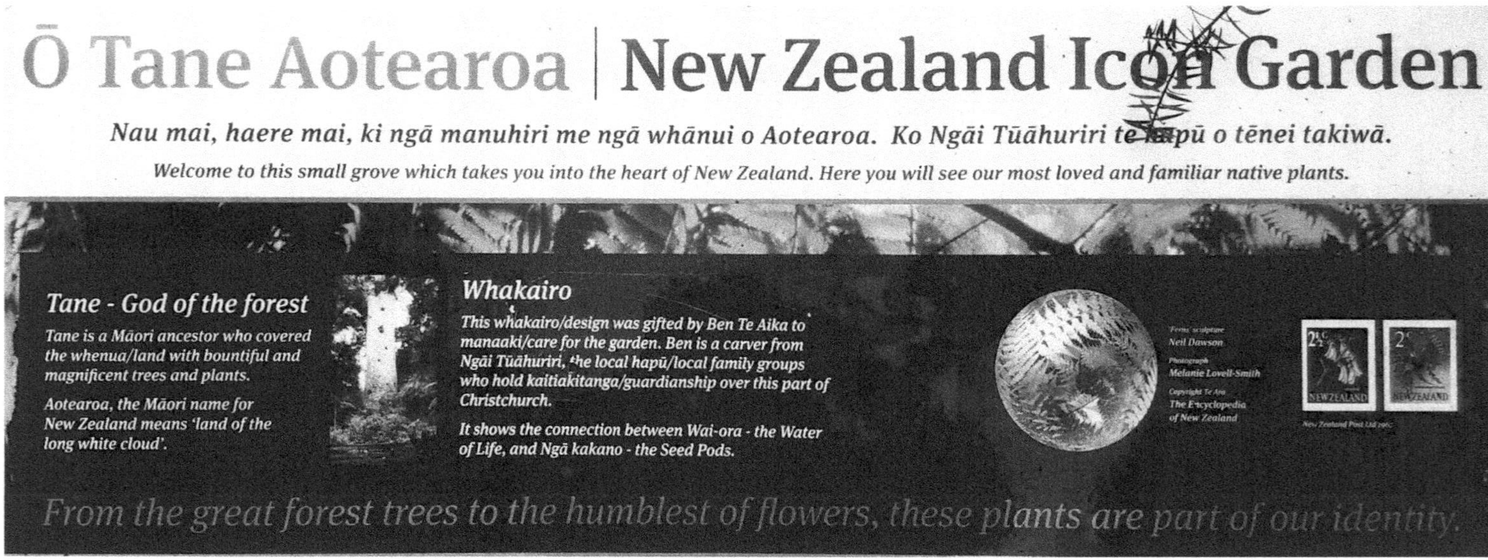

Figure 14.2 Ō Tane Aotearoa, New Zealand Icon Garden, 2016 (B. Hale)

expert on New Zealand flora Leonard Cockayne, provide a more formal display of New Zealand species, planted in beds rather than in the bush-like setting. There is also another area of New Zealand plants, including the national icon, the ponga (silver tree fern), that line the walkway entering the Botanic Gardens along the Natural History Museum. The 1960s–1970s saw the planting of two more native trees on the grand entrance (in 1967, a totara and in 1972, a rimu). More recently, the 2007 Botanic Gardens management plan has placed more emphasis on both New Zealand flora and on Māori perspectives on, and uses of, the natural world stating: "the Gardens are a significant and important area for the support of native New Zealand wildlife (3)." However, as the 2007 plan points out, the individual specimens are planted too close together and the present collection brings together species from a variety of diverse habitats. Thus, while the species represented in the garden are native, their environmental context is completely constructed. In 2014, the gardens added a Visitor Center, which strives to make good on several of the goals from the 2007 management plan, such as highlighting local and regional flora, as well as the traditional Maori use of plants. A permanent exhibit in the Visitor Center provides visitors with an overview of important habitats in the Christchurch region, presents traditional Māori uses of plants such as harakeke (New Zealand flax) and touches upon some of the challenges the region faces in dealing with introduced species, such as gorse.

While the above transformations to the Botanic Gardens indicate an increasing, and important, respect for New Zealand's pre-colonial natural heritage, they are also problematic examples of a larger cultural quest to reinvent New Zealand as a postcolonial nation-state at peace with its colonial past.[67] Such attempts, warn critics, use nature as the "supposedly neutral terrain on which postcolonial New Zealand constructs a shared sense of national community" and have come to represent an emergent form of eco-nationalism among descendants of European colonists.[68] While renewed interest in 'native' heritage, both natural and cultural, is critical to righting colonial wrongs and reinventing postcolonial nations, we must also be wary of such constructions as they are inevitably patently ideological, ecologically impossible (nature is always changing) and they engender understandings of nature as mere representation rather than an active subject.[69]

Riccarton Bush: Building a Wall, New Zealand Style

Riccarton Bush also bears the mark of postcolonial shifts, most notably in attempts to recreate a pre-colonial 'native' landscape within the park. As noted in the introduction to this volume, the 1960–1970s were characterized by the proliferation of the ecological sciences. The subfield of invasion biology/ecology was first articulated in this period with the publication of Charles Elton's (1958) *The Ecology of Invasions by Animals and Plants*[70] and Baker and Stebbins' (1965) *The Genetics of Colonizing Species*.[71] The 1980s saw the development of conservation biology and the birth of restoration ecology each committed to the conservation of native communities.[72] The discourse of 'native,' 'non-native' and 'invasive' species evolved in this period. The growing canon of ecological knowledge of native and non-native species and communities nicely suited the political/cultural needs of that time, specifically in postcolonial nations where supporting 'native' nature might be used or seen as a "postcolonial salve"[73] for lingering colonial transgressions. Of course, the focus on native species and communities was also an important recognition of, and attempt to right, the damage done to ecosystems during the colonial period. Postcolonial transformations in Riccarton Bush mirror these trends.

The gradual implementation of Deans' original wish to conserve the remaining patch of bush took a decidedly ecological approach beginning in the 1970s.[74] At that time, the Trust Board managing the property closed paths to the public,[75] began to actively remove non-native plant species, control the populations of problematic small mammals and replant native plant species, sourcing seeds directly from Riccarton Bush. One of the key issues facing restoration ecologists in New Zealand is the proliferation of non-native, small mammals (e.g. rats, stoats and brushtail possums), introduced in the colonial period. Such mammals have acclimatized well and are problematic inasmuch as they feast on native bird species and impede plant regeneration by overbrowsing vegetation. In addition to poisoning and trapping small 'invasive' mammals, New Zealand began experimenting in the 1970s with pest-exclusion fences specifically designed to keep out the small mammals,[76] thereby reducing the need to control the predators with poisons. In 2004, Riccarton saw the construction of a predator-proof fence that encloses the 7.7 ha bush area containing kahikatea trees over 600 years old.[77] Visitors are permitted within the bush on marked trails that include information about native species therein as well as the historical uses of Pūtaringamotu/Riccarton Bush by both Māori and British settlers. Although the reserve suffers from its geographical location (an 'island' in an urban/suburban sea) and has lost 39 of its 106 native plant species since 1870,[78] it does offer a glimpse of the pre-colonial/European Bush in this region of New Zealand.[79]

The reintroduction of native plant species throughout the Christchurch area is not limited to Riccarton Bush. Since the 1990s native plants have been planted in large numbers throughout Christchurch's green spaces.[80] Recent research in the area also suggests that natural processes (birds and wind) are dispersing native plant species from Riccarton Bush. Doody et al.'s 2010 study revealed that many native species, typically dispersed by birds, are taking root outside the park's fence in people's lawns and gardens; though they also discovered that many homeowners had removed or transplanted them either because they did not know what they were or preferred what was already growing on their properties. In spite of such reintroductions, recent studies reveal that 'European' lawns still make up over two-thirds of Christchurch open spaces[81] and a mere 17% of street and park trees are native.[82] According to Morgenroth and Armstrong, the recent 2010 and 2011 Canterbury earthquakes appear to have had positive consequences for native restoration in the area.[83] While the tragic quakes devastated the region, closing Christchurch historic city center for over two years, native trees often fared better than European transplants. What is more, the designation of a "red zone" of residential housing along the Avon River that was condemned due to liquefaction of its soils and high risk for future destruction in future earthquakes has inspired a management plan that privileges native species. Though its long-term effects remain to be seen, this disaster has presented an opportunity to revisit and reimagine what nature in Christchurch could be.

The desire to recreate a pre-colonial New Zealand nature is as yet an ecological dream and one that, many assert, will never come true. As elsewhere, reintroductions are often both controversial and complicated. Controversial in the sense that human inhabitants have become used to life without such species (e.g. the wolf in Nordic countries) and complicated in the sense that mere reintroduction does not guarantee success or the desired outcome. Likewise, eradication efforts are viewed by some as both futile and potentially risk-laden. Given the near impossibility of eradicating a species, such efforts are viewed by some as time and money ill spent, and in some cases, such efforts brush over the complexities of species effects; namely, many non-natives are not invasive,[84] some natives are invasive,[85] and some non-natives enrich rather than degrade ecosystems.[86] Some may even perpetuate xenophobic attitudes to non-native outsiders[87] and/or celebrate indigenous purism.[88] What is more, in attempting to restore a paradise lost, some restoration efforts effectively create native zoos that attempt to freeze nature in a pre-human time, rather than help it to adapt to existing conditions. Novel approaches with non-native species are turning to the language of resilience; focusing more on species functions and relationships within ecosystems rather than on species' origins and ultimately accepting the interstitial status of many species on this ever-changing planet.[89]

Postcolonial transformations to Christchurch Botanic Gardens and Riccarton Bush reflect and project contemporary ideas about what constitutes nature/the natural as well as concerns about life on a rapidly urbanizing planet. As case studies, they impart valuable lessons about what has happened to nature in what has become known as the Anthropocene. In New Zealand, landscape composition has shifted from completely nature-driven to human-dominated in rapid fire beginning only 1,000 years ago with the Māori and kicking into high gear 200 years ago with the Europeans. It has only been in the last 50 years that we have begun to observe and understand the consequences of such drastic changes. In an urban context, such as Christchurch, botanical gardens and nature reserves are important interstitial sites where these changes are mediated and made manifest. Both the Christchurch Botanic Gardens and Riccarton Bush offer interesting insights into what a postcolonial and novel urban ecosystem might and might not be. It is clear in these two specific cases that managing and restoring "nativeness" remains a central theme in ecological efforts though to different effect. Indeed, in the last few decades, the recognition and promotion of native species has grown in New Zealand and expanded into landscape design, at the level of private gardens and city planning, and the design of cityscapes where native species can thrive. A recent paper offers the development of 'mosaic landscapes' in urban design that include historical (i.e. system reflects a pre-impact natural state), hybrid (i.e. altered systems that maintain some historical aspects where restoration could take place) and novel systems (i.e. system so heavily altered that no restoration is feasible) as essential elements in the regional conservation of biodiversity and promotion of broad management goals.[90] While the present management of both Christchurch Botanic Gardens and Riccarton Bush privilege the native (at least discursively) in ways that must be met with caution, management does in both cases appear to be in line with the framework proposed by Hobbs and others and reflect an awareness of the evolving *nature* of nature in the Anthropocene.

Notes

1 Elizabeth Kolbert, "The Big Kill: New Zealand's Crusade to Rid Itself of Animals," *The New Yorker*, 2015.
2 Kolbert, "The Big Kill."
3 David Quammen cited in Kolbert, 2014.
4 Franklin Ginn, "Colonial Transformations: Nature, Progress and Science in the Christchurch Botanic Gardens," *New Zealand Geographer* 65 (2009), 36.

5 Franklin Ginn, "Extension, Subversion, Containment: Eco-nationalism and (Post)colonial Nature in Aotearoa New Zealand," *Trans Inst Geogr* (2008).
6 William Mark Adams and Martin Mulligan *Decolonizing Nature: Strategies for Conservation in a Post-colonial Era* (London: Earthscan Publications, 2004), 1.
7 Tomaz Mastnak, Julia Elyachar and Tom Boellstorff, "Botanical Decolonization: Rethinking Native Plants," *Environment and Planning D: Society and Space* 32 (2014), 365.
8 See for example Edward Said, *Orientalism* (New York: Pantheon, 1978), Benedict Anderson, *Imagined Communities: Reflections on the Origin and Spread of Nationalism* (London: Verso, 1983) and Homi K. Bhabha *The Location of Culture* (London: Routledge, 1994).
9 For example, Peter Kalliney, *Commonwealth of Letters: British Literary Culture and the Emergence of Postcolonial Aesthetics* (Oxford: Oxford University Press, 2013) and John Marx, *The Modernist Novel and the Decline of Empire* (Cambridge: Cambridge University Press, 2005).
10 Adams and Mulligan, *Decolonizing Nature*, 23.
11 Ibid.
12 Londa Schiebinger and Claudia Swan *Colonial Botany: Science, Commerce and Politics in the Early Modern World* (Philadelphia: University of Pennsylvania Press, 2005) and Adams and Mulligan, *Decolonizing Nature*.
13 Mary Louise Pratt, *Imperial Eyes: Travel Writing and Transculturation* (London: Routledge, 1992); Richard H. Grove, *Green Imperialism: Colonial Expansion, Tropical Island Edens and the Origins of Environmentalism* (Cambridge: Cambridge University Press, 1995).
14 Tomaz Mastnak, Julia Elyachar and Tom Boellstorff, "Botanical decolonization," 364.
15 Lucile H. Brockway, *Science and Colonial Expansion: The Role of the British Royal Botanic Gardens,* (Yale University Press, 2002), 11.
16 William M. Adams, "Nature and the Colonial Mind," in *Decolonizing Nature: Strategies for Conservation in a Post-colonial Era*, eds. William M. Adams and Martin Mulligan (London: Earthscan Publications,2004), 24; Pratt, *Imperial Eyes*, 38.
17 Adams, *Nature*.
18 Adams, *Nature*, 29.
19 Adams, *Nature*.
20 Adams and Mulligan, *Decolonizing,* 7.
21 Adams and Mulligan, *Decolonizing*, 7.
22 Christchurch City Council, *Christchurch Botanic Gardens Management Plan*. (Christchurch: Christchurch City Council, 2007), www.ccc.govt.nz/the-council/plans-strategies-policies-and-bylaws/plans/park-management-plans/botanic-gardens-management-plan/.
23 Ibid.; Colin D. Meurk and David A, Norton, "A Conservation Blueprint for Christchurch," *Canterbury Botanical Society Journal* 22 (1988): 3–15.
24 Meurk and Norton, "A Conservation Blueprint for Christchurch"; Matt S. McGlone and Janet M. Wilmshurst, "Dating Initial Maori Environmental Impact in New Zealand". *Quaternary International* 59 (1999): 5–16.; Jann A. Williams and Carol J. West, "Environmental Weeds in Australia and New Zealand: Issues and Approaches to Management". *Austral Ecology* 25 (2000): 425–444.
25 Jay Mairi and Munir Morad, "The Socioeconomic Dimensions of Biosecurity: The New Zealand Experience," *International Journal of Environmental Studies* 63 (2006), 293–302.
26 Garth R. Harmsworth and Shaun Awatere. "Indigenous Māori Knowledge and Perspectives of Ecosystems". In *Ecosystem Services in New Zealand: Conditions and Trends*., 1st ed., 274–286. (Palmerston North: Manaaki Whenua Press, Landcare Research, 2013).
27 CCC, *Christchurch Botanic Gardens Management Plan*; Louise Beaumont, Dave Pearson Architects Ltd., and Bridget Mosley, *A Conservation Plan for Hagley Park and The Christchurch Botanic Gardens. Volume Three: Botanic Gardens*, (Christchurch: Christchurch City Council, 2013). www.ccc.govt.nz/parks-and-gardens/christchurch-botanic-gardens/about-us/conservation-plan/conservation-plan-volume-3-botanic-gardens/
28 Beaumont et al *A Conservation Plan for Hagley Park and The Christchurch Botanic Gardens. Volume Three: Botanic Gardens*.
29 The British association overseeing the establishment of a colony on this site.
30 Michael M. Roche. "Reactions to Scarcity: The Management of Forest Resources in Nineteenth-Century Canterbury, New Zealand," *Forest & Conservation History* 28 (1984), 82–91.

31 Beaumont et al, *A Conservation Plan for Hagley Park and The Christchurch Botanic Gardens. Volume Three: Botanic Gardens.*
32 CCC, *Christchurch Botanic Gardens Management Plan.*
33 Beaumont et al, *A Conservation Plan for Hagley Park and The Christchurch Botanic Gardens. Volume Three: Botanic Gardens*, 32.
34 Louise Beaumont, David Pearson Architects Ltd, and Bridget Mosley. *A Conservation Plan for Hagley Park and The Christchurch Botanic Gardens. Volume One: History*. (Christchurch: Christchurch City Council, 2013) www.ccc.govt.nz/parks-and-gardens/christchurch-botanic-gardens/about-us/conservation-plan/conservation-plan-volume-1-history/.
35 CCC, *Christchurch Botanic Gardens Management Plan*, Beaumont et al. *A Conservation Plan for Hagley Park and The Christchurch Botanic Gardens. Volume One: History.*
36 CCC, *Christchurch Botanic Gardens Management Plan*
37 Ibid.
38 In 1990, the Acclimatisation Societies in New Zealand underwent a major transformation to focus primarily on managing fish and game resources (both native and introduced) and were recast as *Fish and Game New Zealand*; the regional council with oversight over the Christchurch region is Fish and Game North Canterbury (Walrond 2008).
39 Jann A. Williams and Carol J. West. "Environmental Weeds in Australia and New Zealand: Issues and Approaches to Management". *Austral Ecology* 25 (2000): 425–444.
40 Beaumont et al. *A Conservation Plan for Hagley Park and The Christchurch Botanic Gardens. Volume One: History.*
41 CCC, *Christchurch Botanic Gardens Management Plan.*
42 Ginn, "Colonial Transformations", 43.
43 Beaumont et al, *A Conservation Plan for Hagley Park and The Christchurch Botanic Gardens. Volume Three: Botanic Gardens*, 2.
44 CCC, *Christchurch Botanic Gardens Management Plan.*
45 Beaumont et al, *A Conservation Plan for Hagley Park and The Christchurch Botanic Gardens. Volume Three: Botanic Gardens*, 14.
46 Beaumont et al, *A Conservation Plan for Hagley Park and The Christchurch Botanic Gardens. Volume Three: Botanic Gardens.*
47 Interestingly, the kauri is not native to the South Island and is only naturally found on the North Island.
48 John Dawson and Rob Lucas. *Nature Guide to The New Zealand Forest*. 1st ed. (Auckland, N.Z.: Random House New Zealand, 2000).
49 CCC, *Christchurch Botanic Gardens Management Plan.*
50 Beaumont et al, *A Conservation Plan for Hagley Park and The Christchurch Botanic Gardens. Volume Three: Botanic Gardens.*
51 Roche, "Reactions to Scarcity."
52 Doody, Brendan J. "Riccarton Bush and The Natural and Social Realities of Native Trees in Christchurch, New Zealand". (Masters, Lincoln University 2008); Bruce D. Clarkson and Catherine L. Kirby. "Ecological Restoration in Urban Environments in New Zealand". *Ecological Management & Restoration* 17 (2016), 180–190.
53 Brian Molloy. "History and Management of Riccarton Bush" *The New Zealand Garden Journal (Journal of the Royal New Zealand Institute of Horticulture)* 3 (2000), 13–18.
54 Ibid.
55 Primary remnants have retained a composition of primarily native species.
56 Mark Derby. "Waitangi Tribunal—Te Rōpū Whakamaka—Forming the Waitangi Tribunal, 1970's". (*Teara.Govt.Nz*, 2012) www.teara.govt.nz/en/waitangi-tribunal-te-ropu-whakamana/page-1.
57 Munir Morad and Jay Mairi. "'Freedom's Ramparts on The Sea': The Postcolonial New Zealand Landscape and The Quest for Sustainability". *British Review of New Zealand Studies* 14 (2003), 119–130.
58 Davis, Mark A. "Invasion Biology 1958–2005: The Pursuit of Science and Conservation". In *Conceptual Ecology and Invasion Biology: Reciprocal Approaches to Nature*, 1st ed. (Springer Netherlands, 2006), 35–64.
59 Adams and Mulligan, *Decolonizing*.
60 Rodney Harrison, et al, "Introduction: Heritage, Memory and Modernity," in *The Heritage Reader*, eds. Graham Fairclough, Rodney Harrison, John Schofield, Jr and John H. Jameson (London: Routledge, 2007), 1–2.

61 The World Heritage Convention n.d. UNESCO, Retrieved on March 10, 2017 from http://whc.unesco.org/en/convention/
62 The World Heritage Convention n.d. UNESCO, Retrieved on March 10, 2017 from http://whc.unesco.org/en/convention/
63 The World Heritage Convention n.d. UNESCO, Retrieved on March 10, 2017 from http://whc.unesco.org/en/convention/
64 Cultural Landscapes, n.d.
65 Ginn, "Colonial Transformations."
66 CCC *Christchurch Botanic Gardens Management Plan*; Christchurch City Council. *Christchurch Botanic Gardens Walking Guide*, (Christchurch: Christchurch City Council, 2011) www.ccc.govt.nz/assets/Documents/Parks-Gardens/Christchurch-Botanic-Gardens/BotanicGardensWalkingGuide.pdf.
67 Ginn "Extension, Subversion, Containment."
68 Ibid., 238.
69 Ibid.
70 Charles S. Elton, *The Ecology of Invasions by Animals and Plants*. (London: Methuen, 1958).
71 H. G. Baker and G. Ledyard Stebbins. *The Genetics of Colonizing Species. Proceedings of the First International Union of Biological Sciences Symposium on General Biology* (New York & London: Academic Press, 1965).
72 Davis "Invasion Biology."
73 Ginn, "Colonial Transformations."
74 Clarkson and Kirby, "Ecological Restoration."
75 Molloy, "History and Management of Riccarton Bush."
76 Bruce Burns, John Innes and Tim Day. "The Use and Potential of Pest-Proof Fencing for Ecosystem Restoration and Fauna Conservation in New Zealand". In *Fencing for Conservation: Restriction of Evolutionary Potential or A Riposte to Threatening Processes?* 1st ed. (Springer Science+Business Media, 2012) pp. 65–90.
77 Riccarton House and Bush. "Riccarton Bush—Riccarton House and Bush." (2017) www.riccartonhouse.co.nz/riccarton-bush/.
78 Brendan J. Doody, Jon J. Sullivan, Colin D. Meurk, Glenn H. Stewart, and Harvey C. Perkins. "Urban Realities: The Contribution of Residential Gardens to The Conservation of Urban Forest Remnants". *Biodiversity and Conservation* 19 (2010), 1385–1400.
79 e.g. Meurk and Given, "A Conservation Blueprint for Christchurch; Molloy "History and Management of Riccarton Bush"; Glenn H. Stewart, Maria E. Ignatieva, Colin D. Meurk, Hannah Buckley, Benjamin Horne, and Toni Braddick. "Urban Biotopes of Aotearoa New Zealand (URBANZ) (I): Composition and Diversity of Temperate Urban Lawns in Christchurch," *Urban Ecosystems* 12 (2009), 233–248; Doody et al. "Urban Realities.
80 Colin D. Meurk. "Threats to Native Plants and Niches for Survival in New Zealand Cultural Landscapes," *Canterbury Botanical Society Journal* 36 (2002), 18–25.
81 G.H. Stewart, C.D. Meurk, M.E. Ignatieva, H.L. Buckley, A. Magueur, B.S. Case, M. Hudson, and M. Parker. "Urban Biotopes of Aotearoa New Zealand (URBANZ) II: Floristics, Biodiversity and Conservation Values of Urban Residential and Public Woodlands, Christchurch,". *Urban Forestry & Urban Greening* 8 (2009), 149–162.
82 Glenn H. Stewart, Maria E. Ignatieva, Colin D. Meurk, and Richard D. Earl, "The Re-Emergence of Indigenous Forest in an Urban Environment, Christchurch, New Zealand". *Urban Forestry & Urban Greening* 2, no. 3 (2004), 149–158.
83 Justin Morgenroth and Tony Armstrong. "The Impact of Significant Earthquakes on Christchurch, New Zealand's Urban Forest," *Urban Forestry & Urban Greening* 11 (2012), 383–389.
84 David M. Richardson, Petr Pyšek, and James T. Carlton, "A Compendium of Essential Concepts and Terminology in Invasion Ecology," *Fifty Years of Invasion Ecology. The Legacy of Charles Elton* (2011), 409–420.
85 Joy B. Zedler, and Suzanne Kercher, "Causes and Consequences of Invasive Plants in Wetlands: Opportunities, Opportunists, and Outcomes," *Critical Reviews in Plant Sciences* 23, no. 5 (2004), 431–452.
86 Martin A. Schlaepfer, Dov F. Sax, and Julian D. Olden, "The Potential Conservation Value of Non-Native Species," *Conservation Biology* 25, no. 3 (2011), 428–437.

87 Gert Gröning, Gert, and Joachim Wolschke-Bulmahn. "The Native Plant Enthusiasm: Ecological Panacea or Xenophobia?" *Arnoldia* 62, no. 4 (2004), 20–28.
88 Lesley Head and Pat Muir. "Nativeness, Invasiveness, and Nation in Australian Plants," *Geographical Review* (2004), 199–217.
89 e.g. Susan Cordell, Rebecca Ostertag, Jené Michaud, and Laura Warman, "Quandaries of a Decade-Long Restoration Experiment Trying to Reduce Invasive Species: Beat Them, Join Them, Give up, or Start Over?." *Restoration Ecology* (2016), 139–144.
90 Richard J. Hobbs, Eric Higgs, Carol M. Hall, Peter Bridgewater, F. Stuart Chapin, Erle C. Ellis, John J. Ewel et al., "Managing the Whole Landscape: Historical, Hybrid, and Novel Ecosystems," *Frontiers in Ecology and the Environment* 12, no. 10 (2014), 557–564.

References

Adams, William M. "Nature and the Colonial Mind." in William M. Adams and Martin Mulligan (eds.), *Decolonizing Nature: Strategies for Conservation in a Post-Colonial Era*, 16–50. London: Earthscan Publications, 2004.

Adams, William M., and Martin Mulligan. *Decolonizing Nature: Strategies for Conservation in a Post-Colonial Era*. London: Earthscan Publications, 2004.

Baker, H.G. and G. Ledyard Stebbins. *The Genetics of Colonizing Species*. Proceedings of the First International Union of Biological Sciences Symposium on General Biology, New York & London: Academic Press, 1965.

Beaumont, Louise, Dave Pearson Architects Ltd., and Bridget Mosley. *A Conservation Plan for Hagley Park and the Christchurch Botanic Gardens. Volume Three: Botanic Gardens*, 2013a. Christchurch: Christchurch City Council. retrieved from www.ccc.govt.nz/parks-and-gardens/christchurch-botanic-gardens/about-us/conservation-plan/conservation-plan-volume-3-botanic-gardens/.

Beaumont, Louise, David Pearson Architects Ltd, and Bridget Mosley. *A Conservation Plan for Hagley Park and the Christchurch Botanic Gardens. Volume One: History*. Christchurch: Christchurch City Council, 2013b. retrieved from www.ccc.govt.nz/parks-and-gardens/christchurch-botanic-gardens/about-us/conservation-plan/conservation-plan-volume-1-history/.

Belich, James. *Make Peoples: A History of the New Zealanders from Polynesian Settlement to the End of the Nineteenth Century*. Honolulu: University of Hawaii Press, 1996.

Brockway, Lucile H. *Science and Colonial Expansion: The Role of the British Royal Botanic Gardens*. New Haven: Yale University Press, 2002.

Boffa Miskell Limited. *RESIDENTIAL RED ZONE: Vegetation Retention Methodology—Version 2.0*. Christchurch: Land Information New Zealand, 2012. retrieved from www.linz.govt.nz/system/files_force/media/pages-attachments/crown-property_rrz-vegetation-retention_2014%2007%2029.pdf.

Burns, Bruce, John Innes, and Tim Day. "The Use and Potential of Pest-Proof Fencing for Ecosystem Restoration and Fauna Conservation in New Zealand." in *Fencing for Conservation: Restriction of Evolutionary Potential or a Riposte to Threatening Processes?* 1st ed., 65–90. Berlin: Springer Science+Business Media, 2012.

[CCC] Christchurch City Council. *Christchurch Botanic Gardens Management Plan*. Christchurch: Christchurch City Council, 2007. retrieved from www.ccc.govt.nz/the-council/plans-strategies-policies-and-bylaws/plans/park-management-plans/botanic-gardens-management-plan/.

[CCC] Christchurch City Council. *Christchurch Botanic Gardens Walking Guide*. Christchurch: Christchurch City Council, 2011. retrieved from www.ccc.govt.nz/assets/Documents/Parks-Gardens/Christchurch-Botanic-Gardens/BotanicGardensWalkingGuide.pdf.

Clarkson, Bruce D., and Catherine L. Kirby. "Ecological Restoration in Urban Environments in New Zealand." *Ecological Management & Restoration* 17, no. 3 (2016): 180–190. doi:10.1111/emr.12229.

Coombes, Brad. "Postcolonial Conservation and Kieke Harvest at Morere New Zealand: Abstracting Indigenous Knowledge From Indigenous Polities." *Geographical Research* 45 (2007): 186–193.

Cordell, Susan, Rebecca Ostertag, Jené Michaud, and Laura Warman. "Quandaries of a Decade-Long Restoration Experiment Trying to Reduce Invasive Species: Beat Them, Join Them, Give Up, Or Start Over?". *Restoration Ecology* 24, no. 2 (2016): 139–144. doi:10.1111/rec.12321.

Davis, Mark A. "Invasion Biology 1958–2005: The Pursuit of Science and Conservation." in *Conceptual Ecology and Invasion Biology: Reciprocal Approaches to Nature*, 1st ed., 35–64. Netherlands: Springer, 2006.

Dawson, John, and Rob Lucas. *Nature Guide to The New Zealand Forest.*, 1st ed. Auckland, NZ: Random House New Zealand, 2000.

Department of Conservation. *Ngāi Tahu Taonga Plant Species*. Christchurch: Department of Conservation/Te Papa Atawhai, 2006.

Derby, Mark. "Waitangi Tribunal—Te Rōpū Whakamaka—Forming the Waitangi Tribunal, 1970'S." *Teara.Govt.Nz*, 2012. retrieved from www.teara.govt.nz/en/waitangi-tribunal-te-ropu-whakamana/page-1.

Doody, Brendan J. "Riccarton Bush and the Natural and Social Realities of Native Trees in Christchurch, New Zealand." Masters, Lincoln University, 2008.

Doody, Brendan J., Jon J. Sullivan, Colin D. Meurk, Glenn H. Stewart, and Harvey C. Perkins. "Urban Realities: The Contribution of Residential Gardens to the Conservation of Urban Forest Remnants." *Biodiversity and Conservation* 19, no. 5 (2010): 1385–1400. doi:10.1007/s10531-009-9768-2.

Drayton, Richard. *Nature's Government: Science, Imperial Britain and the 'Improvement' of the World*. New Haven: Yale University Press, 2000.

Dumplemann, Sonja. *A Cultural History of Gardens in the Age of Empire*. London: Bloomsbury Academic, 2016.

Ginn, Franklin. "Extension, Subversion, Containment: Eco-Nationalism and (Post)Colonial Nature in Aotearoa New Zealand." *Trans Inst Geogr* (2008): 335, 353.

Ginn, Franklin. "Colonial Transformations: Nature, Progress and Science in the Christchurch Botanic Gardens." *New Zealand Geographer* 65 (2009): 35–47.

Given, David, and Colin Meurk. "Biodiversity of the Urban Environment: The Importance of Indigenous Species and the Role Urban Environments Can Play in Their Preservation." in *Urban Biodiversity and Ecology as a Basis for Holistic Planning and Design*, 22–33. Christchurch: Wickliffe Press Ltd, 2000.

Gröning, Gert, and Joachim Wolschke-Bulmahn. "The Native Plant Enthusiasm: Ecological Panacea or Xenophobia?" *Arnoldia* 62, no. 4 (2004): 20–28.

Grove, Richard H. *Green Imperialism: Colonial Expansion, Tropical Island Edens and the Origins of Environmentalism*. Cambridge: Cambridge University Press, 1995.

Harmsworth, Garth R., and Shaun Awatere. "Indigenous Māori Knowledge and Perspectives of Ecosystems." in *Ecosystem Services in New Zealand: Conditions and Trends*, 1st ed., 274–286. Palmerston North: Manaaki Whenua Press, Landcare Research, 2013.

Harrison, R. et al. "Introduction: Heritage, Memory and Modernity." in Graham Fairclough, Rodney Harrison, John Schofield, Jr and John H. Jameson (eds.), *The Heritage Reader*. London: Routledge, 2007.

Head, Lesley, and Pat Muir. "Nativeness, Invasiveness, and Nation in Australian Plants." *Geographical Review* (2004): 199–217.

Jay, M., and M. Morad. "The Socioeconomic Dimensions of Biosecurity: The New Zealand Experience." *International Journal of Environmental Studies* 63, no. 3 (2006): 293–302.

Kalliney, P. *Commonwealth of Letters: British Literary Culture and the Emergence of Postcolonial Aesthetics*. Oxford: Oxford University Press, 2013.

Marx, J. *The Modernist Novel and the Decline of Empire*. Cambridge: Cambridge University Press, 2005.

Mastnak, Tomaz, Julia Elyachar, and Tom Boellstorff. "Botanical Decolonization: Rethinking Native Plants." *Environment and Planning D: Society and Space* 32 (2014): 363–380.

McGlone, Matt S., and Janet M. Wilmshurst. "Dating Initial Maori Environmental Impact in New Zealand." *Quaternary International* 59, no. 1 (1999): 5–16. doi:10.1016/s1040-6182(98)00067-6.

Meurk, Colin D., and David A. Norton. "A Conservation Blueprint for Christchurch." *Canterbury Botanical Society Journal* 22 (2017): 3–15.

Meurk, Colin D. "Threats to Native Plants and Niches for Survival in New Zealand Cultural Landscapes." *Canterbury Botanical Society Journal* 36 (2002): 18–25.

Morad, Munir, and Mairi Jay. "'Freedom's Ramparts on the Sea': The Postcolonial New Zealand Landscape and the Quest for Sustainability." *British Review of New Zealand Studies* 14 (2003): 119–130.

Morgenroth, Justin, and Tony Armstrong. "The Impact of Significant Earthquakes on Christchurch, New Zealand's Urban Forest." *Urban Forestry & Urban Greening* 11, no. 4 (2012): 383–389. doi:10.1016/j.ufug.2012.06.003.

Quigley, Mark C., Matthew W. Hughes, Brendon A. Bradley, Sjoerd van Ballegooy, Catherine Reid, Justin Morgenroth, Travis Horton, Brendan Duffy, and Jarg R. Pettinga. "The 2010–2011 Canterbury Earthquake Sequence: Environmental Effects, Seismic Triggering Thresholds and Geologic Legacy." *Tectonophysics* 672–673 (2016): 228–274. doi:10.1016/j.tecto.2016.01.044.

UNESCO. *The World Heritage Convention: Strategic Objectives, the 'Five Cs'*. retrieved on March 10, 2017 from http://whc.unesco.org/en/convention/.

Pratt, Mary Louise. *Imperial Eyes: Travel Writing and Transculturation*. London: Routledge, 1992.

Riccarton House and Bush. "Riccarton Bush—Riccarton House and Bush." 2017. retrieved from www.riccartonhouse.co.nz/riccarton-bush/.

Roche, Michael M. "Reactions to Scarcity: The Management of Forest Resources in Nineteenth-Century Canterbury, New Zealand." *Forest & Conservation History* 28, no. 2 (1984): 82–91. doi:10.2307/4004774.

Schiebinger, Londa, and Claudia Swan. *Colonial Botany: Science, Commerce and Politics in the Early Modern World*. Philadelphia: University of Pennsylvania Press, 2005.

Stewart, G.H., C.D. Meurk, M.E. Ignatieva, H.L. Buckley, A. Magueur, B.S. Case, M. Hudson, and M. Parker. "Urban Biotopes of Aotearoa New Zealand (URBANZ) II: Floristics, Biodiversity and Conservation Values of Urban Residential and Public Woodlands, Christchurch." *Urban Forestry & Urban Greening* 8, no. 3 (2009): 149–162. doi:10.1016/j.ufug.2009.06.004.

Stewart, Glenn H., Maria E. Ignatieva, Colin D. Meurk, and Richard D. Earl. "The Re-Emergence of Indigenous Forest in an Urban Environment, Christchurch, New Zealand." *Urban Forestry & Urban Greening* 2, no. 3 (2004): 149–158. doi:10.1078/1618–8667-00031.

Stewart, Glenn H., Maria E. Ignatieva, Colin D. Meurk, Hannah Buckley, Benjamin Horne, and Toni Braddick. "Urban Biotopes of Aotearoa New Zealand (URBANZ) (I): Composition and Diversity of Temperate Urban Lawns in Christchurch." *Urban Ecosystems* 12, no. 3 (2009): 233–248. doi:10.1007/s11252-009-0098-7.

Walrond, Carl. "Acclimatisation—Fish and Game Councils.". *Te Ara*, 2008. retrieved from www.teara.govt.nz/en/acclimatisation.

Williams, Jann A., and Carol J. West. "Environmental Weeds in Australia and New Zealand: Issues and Approaches to Management." *Austral Ecology* 25, no. 5 (2000): 425–444. doi:10.1046/j.1442-9993.2000.01081.x.

Contributors

Gabriel N. Gee is Associate Professor in Art History at Franklin University, Switzerland. Recent publications include *Art in the North of England, 1979–2008* (Routledge—an Ashgate book 2017). His current research interests include 20th-century British and Irish art, the changing representations and imaginaries of port cities in the second half of the 20th century, as well as interconnected global histories, with a particular interest in urban and architectural representation. He is a co-founder of the TETI group, for *Textures and Experiences of Transindustriality* (www.tetigroup.org).

Lori M. Gibbs is an architectural designer and PhD candidate in Architecture at the University of Pennsylvania. Her dissertation examines methods of survey documentation and drawings that craft architectural histories in a cross-cultural context between 1920–1940. Previously she taught research seminars and lecture courses in the Faculty of Architecture and Division of Landscape Architecture at the University of Hong Kong. She earned an MA in Histories and Theories of Architecture at the Architectural Association in London, and a B.Arch from Pratt Institute. She is a LEED AP and maintains an architectural practice in New York with architect Luis Silva da Costa.

Brack W. Hale is Associate Professor of Biology and Environmental Science at Franklin University Switzerland where he has worked since 2006; he is also the co-director of the Center for Sustainability Initiatives at Franklin. His research interests have revolved around how humans interact with and impact the natural world and he has published numerous articles in the fields of biofuels, ecology and management of riparian areas, invasive species, natural areas conservation, sustainable travel and tourism, and waste management.

Ayala Levin is Assistant Professor in the Department of Art History at Northwestern University. Levin is an architectural historian whose research is concerned with north-south and south-south architectural knowledge exchange, with a focus on building and urban planning projects in post-independence African states. She is currently completing a monograph on the export of Israeli architectural and planning models to Sierra Leone, Nigeria, Ethiopia and the Ivory Coast in the 1960s–1970s. Her next book-length project will explore the emergence of environmental planning in the US in conjunction with the work of American architects, planners and landscape architects in Africa from the mid-1950s to the late 1970s. Levin completed her PhD at Columbia University Graduate School of Architecture, Planning and Preservation. Before joining Northwestern University, she was a fellow at the Princeton-Mellon Initiative in Architecture, Urbanism and the Humanities, a post-doctoral researcher in the ERC project "Apartheid: The Global Itinerary," and a Visiting Assistant Professor at Pratt Institute.

Maria João Matos is a researcher based in Lisbon. She obtained a diploma in Architecture, from the Faculty of Architecture, Technical University of Lisbon (UTL), Portugal. She holds a PhD in Architecture and Landscape from both University of Beira Interior, Portugal and University Paris 8, France. She received a Master's degree in "City, Territory and Urban Renewal" from University Institute of Lisbon (ISCTE). She worked as an Assistant Professor at the Architecture Department of Lusófona University, Lisbon – Lusófona University of Humanities and Technologies and as a researcher in the areas of Urban Design and Landscape, at Research Centre in Architecture, Urbansim and Design and Laboratory for Architecture and Art projects/Lusófona University of Humanities and Technologies.

Conor McFeely is an artist who lives and works in Derry N. Ireland. His work incorporates a wide range of processes, from the ready—made to sculpture and installation, as well as photography, video and audio. A fracturing and manipulation of 'material' in the service of finding new relationships is a chief characteristic of his practice. Often conceived as multi-layered in terms of their reading, many works have been driven by ruminations on the nature of individual freewill, choice and autonomy. Contexts and source material reflect interests in a history of counter culture, literature and social contexts. His recent installation project *Pioneer* points to episodes of altered states in relation to religious rapture and sexual ecstasy, as well as drug-induced experiences and interruptions of consciousness.

Tuula Närhinen is a visual artist based in Helsinki, Finland. Her works explore the pictorial agency of natural phenomena such as water and wind. Re-adapting instruments derived from natural sciences, she has developed methods for letting trees trace the shape of wind on their branches and found techniques that the enable the waves of the sea to inscribe themselves on paper. Her installations showcase the DIY instruments implicated, encouraging the spectator to participate in the re-presentation of an event. Närhinen holds a Doctorate of Fine Arts (DFA) from the University of the Arts Helsinki. She is a graduate of the Finnish Academy of Fine Arts in Helsinki (MFA), and the Helsinki University of Technology (M.Sc. in Architecture).

Uriel Orlow is an artist who lives and works in London and Zurich. His practice is research-based and emphasizes artistic *processes*. His work takes various forms using different mediums, including film, photography, drawing, lecture performances and modular, multi-media installations that focus upon specific locations and micro-histories, bringing different image regimes and narrative modes into correspondence. Orlow is particularly concerned with the spatial manifestations of memory, blind spots of representation and forms of haunting.

Paolo Perulli is Professor of Economic Sociology at the University of Eastern Piedmont (Italy), and Professor of Urban Governance, Architecture Academy of Mendrisio, (Switzerland). Paolo has been a visiting scholar at MIT and a Professeur invité at the Université de Paris Sud. His recent publications include: *Visioni di città*

(Einaudi 2009), *Il dio Contratto* (Einaudi 2012), *Terra mobile* (Einaudi 2014) and *The Urban Contract* (Routledge 2017).

Eliana Sousa Santos is an architect, postdoctoral researcher and Assistant Professor of Architecture in the Department of Architecture at ECATI ULHT, Lisbon. Sousa Santos has a degree in architecture from the Technical University of Lisbon, a master degree from University of Coimbra and a PhD from the University of London. She was a visiting postdoctoral research fellow at Yale University in 2013/14. Eliana curated the exhibition *The Shape of Plain* at the Gulbenkian Museum, an associated project of the Lisbon Architecture Triennial 2016. She is currently working on the project "George Kubler's Shape of Time: The Historiographical effect of Portuguese Plain Architecture in Post- revolutionary Portugal" at CES, University of Coimbra.

Giuliano Sergio is Professor of Art History at the Academy of Fine Arts in Urbino, Italy. Sergio gained his PhD from the Université Paris X and the Università di Roma "La Sapienza." His publications include *Information document oeuvre. Parcours de la photographie en Italie dans les années soixante et soixante-dix* (Presses universitaires de Paris Ouest, 2015), *Atlante degli archivi fotografici e audiovisivi italiani digitalizzati* (Marsilio, 2015) and *Ugo Mulas. Vitalità del negativo* (Johan & Levi, 2010). He has curated a number of exhibitions in recent years, including *The Art of the Feminine* (Villa Pignatelli, Naples 2017), *Paolo Gioli. Corpi favoriti dalla notte* (IIC Madrid 2017), the itinerant installation *La montagne de Venise* by Yona Friedma with Jean-Baptiste Decavèle (Querini Stampalia/IUAV, Venice 2016), *Ugo Mulas La photographie* (Fondation Henri Cartier Bresson, Paris 2016), *Paesaggi d'aria. Luigi Ghirri e Yona Friedman/Jean-Baptiste Decavèle* (Querini Stampalia, Venice 2015), *Luigi Ghirri. Pensare per immagini* (MAXXI, Rome 2013 and Instituto Moreira Salles, San Paolo and Rio de Janeiro 2013–2014).

Michelle L. Stefano is a Folklife Specialist at the American Folklife Center, Library of Congress (Washington, DC). She earned her BA in art history (Brown University, 2000), MA in international museum studies (Gothenburg University, Sweden, 2004) and PhD in 2010 from the International Centre for Cultural and Heritage Studies at Newcastle University (UK). From 2011–2016, Stefano worked for Maryland Traditions, the folklife program of Maryland, of which she was its Co-director from 2015–2016. From 2012–2016, she led the partnership between Maryland Traditions and the University of Maryland, Baltimore County, where she was Visiting Assistant Professor in American Studies. She co-edited *The Routledge Companion to Intangible Cultural Heritage* (2016), *Engaging Heritage, Engaging Communities* (Boydell & Brewer, 2016), and *Safeguarding Intangible Cultural Heritage* (Boydell & Brewer, 2012).

Worrasit Tantinipankul is a registered landscape architect and architect in Thailand. He received his PhD in Historic Preservation Planning from Cornell University in 2006 and Master of Landscape Architecture from the University of Pennsylvania in 1998. He teaches architectural design, urban planning and history of Thai architecture at the School of Architecture and Design (International Program), King Mongkut's University of Technology Thonburi (KMUTT). He received grants from the National Research Council of Thailand in 2013 and the US Ambassador Fund for

Cultural Preservation in 2014 for his continuing research on the vernacular documentation and conservation of the old historic canal communities of Thonburi, the western district of Bangkok.

Alison Vogelaar is an Associate Professor of Communication and Media Studies at Franklin University Switzerland where she teaches courses that variously explore the relationship between media, communication, and power. Her research interests include environmental and ecological discourses, advocacy and activism, maker and craft culture and educational travel. She has recently co-edited, with Brack Hale and Alexandra Peat, another environmentally themed volume, *The Discourses of Environmental Collapse: Imagining the End* (Routledge 2018).

Jundan Zhang is a post-doc researcher within the Department of Geography and Economic History at Umeå University, Sweden. Her current research focuses on relationships between environmental discourses and environmental subjectivities in the context of tourism in peripheral areas, in particular how individuals translate and appropriate broad ideas on 'nature' into their exchanges with other human and non-human actors through tourism activities, and how these translations and appropriations in turn contribute to the discourses of environment.

Index

For Product Safety Concerns and Information please contact our EU representative GPSR@taylorandfrancis.com
Taylor & Francis Verlag GmbH, Kaufingerstraße 24, 80331 München, Germany

www.ingramcontent.com/pod-product-compliance
Ingram Content Group UK Ltd.
Pitfield, Milton Keynes, MK11 3LW, UK
UKHW050754180926
475701UK00007B/1040